T0214908

SOCIAL TRANSFORMATION FOR CLIMATE CHANGE

This book argues that social transformation is both necessary and possible if democracies are to respond effectively to the climate crisis without social collapse.

Climate transformation and social transformation are intimately connected. Understanding how to address climate change requires a historical approach both to the climate and to our collective institutions of humanity. Drawing on the works of Karl Polanyi and Thomas Piketty, Nicholas Low traces the course of historic social transformations from Britain, Russia, and Australia to highlight key commonalities: social crisis, the widespread sense by those in power that 'something has to change', the shift in ideology, and the political champions that drove the change. Within its international scope, the book delves deeper into specific instances of inequality and poverty from Britain, the USA, Australia, and the Global South. It shows how these examples are connected with the current climate emergency. Finally, the author draws together all the evidence from past transformations to outline how a new social democratic transformation could generate a better future, creating the social solidarity necessary to cope with the climate crisis.

This book will be of great interest to students and scholars of climate change, environmental politics and policy, political ecology, environmental sociology, and environmental studies more broadly. Its argument is also highly relevant for political actors working towards social and economic transformation.

Nicholas Low is a Professorial Fellow of the University of Melbourne. After qualifying as an architect at the Architectural Association in London, he took his Master of Urban Planning degree at Strathclyde University, Glasgow. Tutored by sociologist, Dr Sean Damer, he found that there was much more to town planning than drawing patterns on maps. At the core of planning lay political doctrines and institutions. Nick then worked as a city planner in the outer London Borough of Hillingdon. After moving to Melbourne University in 1974 as a lecturer in planning, he studied the evolution of cities, planning, and environments and their political contexts – with growing concern for the substitution of the market and politicians for professional planning. In the 1990s turning towards environmental planning, he found that democracies were failing to prepare for an overheating world. In 1997 he convened the first international conference on environmental justice (*Environmental Justice: Global Ethics for the 21st Century*). He has authored, co-authored, or edited 11 books including *Planning, Politics and the State* (sole author, 1991), *Justice, Society and Nature: An Exploration of Political Ecology* (1998 with Brendan Gleeson), *The Green City: Sustainable Homes, Sustainable Suburbs* (2005, with Gleeson, Green, and Radovic), *Transforming Urban Transport: The Ethics, Politics and Practices of Sustainable Mobility* (ed. 2013), and *Being a Planner in Society: For People, Planet, Place* (sole author, 2020). His work is also published in many international refereed journals and book chapters.

Routledge Advances in Climate Change Research

For more information about this series, please visit: www.routledge.com/Routledge-Advances-in-Climate-Change-Research/book-series/RACCR

SOCIAL TRANSFORMATION FOR CLIMATE CHANGE

A New Framework for Democracy

Nicholas Low

Routledge
Taylor & Francis Group

LONDON AND NEW YORK

earthscan
from Routledge

First published 2024
by Routledge
4 Park Square, Milton Park, Abingdon, Oxon OX14 4RN

and by Routledge
605 Third Avenue, New York, NY 10158

Routledge is an imprint of the Taylor & Francis Group, an informa business

British Library Cataloguing-in-Publication Data
A catalogue record for this book is available from the British Library

ISBN: 978-1-032-46530-2 (hbk)
ISBN: 978-1-032-46531-9 (pbk)
ISBN: 978-1-003-38213-3 (ebk)

DOI: 10.4324/9781003382133

Typeset in Times New Roman
by Deanta Global Publishing Services, Chennai, India

CONTENTS

For Torbjörn Caradog Low (born 31/08/2023) and all future generations everywhere.

ACKNOWLEDGEMENTS

This book is the culmination of some 40 years of my thinking and writing about planning, the state, democracy, social justice, and the environment. In this book, I am happy to join the growing ranks of what Friedrich Hayek called the 'second-hand dealers in ideas', but not *his* second-hand dealers who today fill the profession of economic advisors to governments situated in bloated plutocratic consultancies and neoliberal 'think tanks'.

We all stand on the shoulders of the giants of political economy. Hayek positioned himself on the backs of the nineteenth century theorists of brute capitalism who tended to forget the eighteenth-century moral philosophy of Adam Smith.

My giants are more recent, Karl Polanyi, Thomas Piketty, Tony Judt, and many others. I thank Thomas Piketty for permission to reference his charts of the growth of income equality. I also acknowledge most warmly the authors of the *World Inequality Report* 2022 (Lucas Chancel, Thomas Piketty, Emmanuel Saez, and Gabriel Zucman) and of the *Climate Inequality Report* 2023 (Lucas Chancel, Philipp Bothe, and Tancrède Voituriez) whose work I draw on. I cannot emphasise enough how important these accessible online reports are for those who care about the social dimension of the climate crisis[1]. I thank the European Centre for Medium-Range Weather Forecasts for permission to reproduce the image on the cover of this book.

Reading the recent volumes of Piketty made me go back to Polanyi, whose work became submerged during much of the twentieth century. His ideas still struggle for air with the ideologies of the neoliberal regression. My approach to the subject of transformation I term 'critical history' – exemplified in the work of each of the above writers.

While admiring the philosophical scope, empirical investigations, and passion for justice of Marx and Engels, I was never happy in their company. In a

Marxist-Leninist reality, I would have been denounced as bourgeois (which I am) and an enemy of the people (which I am not). I was pleased to discover recently that my great grandfather, Frederick Low, as Liberal MP for Norwich in England, voted for Lloyd George's 'people's budget' which was the first step, revolutionary in its time, towards the democratic socialism of the welfare state and away from rule by hereditary nobility.

Democracy is not the ideological scaffolding of capitalism. Democracy is the struggle to embed markets in structures of social control. Democracy is the necessary social armament with which to contain global heating.

Marxism collapsed in 1989. Its wraith lingers on in the dreams of some activists. But even avowed Marxists like the admirable David Harvey, as I have argued, moved away from orthodox Marxism. Unfortunately, what has replaced Marxism in revolutionary ideology in many countries is Fascism. Which brings us back to Polanyi who saw fascism sprouting from the consequences of liberal dogmas in the early twentieth century.

I am optimistic about the future of democracy simply because so many people all over the world are fighting to defend what they have of it and, when they don't have it, sacrificing their lives to get it. Modern authoritarians have always believed that they are the future, but it doesn't take long for them to perish. The 'Thousand Year Reich' of the Nazis lasted just ten years, Stalinism a little longer. The reign of Vladimir Putin and his ageing ex-KGB cabal will be shorter still.

I am indebted to many people. My wife, Liz Low, has been patient and supportive as every morning I disappear into my 'cave' to read and write, even though the stuff of my work is not what she writes about. When this book was still an idea in gestation, I sent the gist of it to Patricia Mayhew OBE, an eminent British sociologist. Pat has been a good friend since 1966 when she and I were part of a group visiting the University of the North East, Argentina, (under the direction of Professor Brian Thompson of that University) to carry out analysis of data gathered for the Chaco Province. What I got from Pat was stringent and constructive criticism which made me rethink my whole concept and reshape it into the form of this book.

Friends and colleagues have helped with conversation and discussion: Professor Jim Falk, Professor Brendan Gleeson, Mr Roger Taylor. The immense scope of Jim Falk's book with Joseph Camilleri *Worlds in Transition: Evolving Government across a Stressed Planet* inspired me to think big. Roger's convening of a series of wide-ranging and far-sighted Melbourne Forums on the environment, transport and planning, and chairing the citizen think tank 'Transport for Melbourne' have alerted us to the problems ahead. Professor Robyn Eckersley put me on the trail of the concept of transformation.

Finally, and importantly, I warmly thank Annabelle Harris, Jyotsna Gurung and Pradiksha Dharsini of Routledge/Taylor and Francis for their support and guidance. I thank the anonymous reviewers who took the trouble to examine my proposal and offer suggestions for improvement, and the whole editorial team at Routledge including proof readers and copy editors.

I'm also indebted to the Dean and Faculty of Architecture, Building and Planning at the University of Melbourne where I was allowed to continue as a Professorial Fellow after I retired.

Note

1 (*World Inequality Report*: https://wir2022.wid.world/www-site/uploads/2021/12/WorldInequalityReport2022_Full_Report.pdf; *Climate Inequality Report*: https://wid.world/wp-content/uploads/2023/01/CBV2023-ClimateInequalityReport-2.pdf) https://wir2022.wid.world/www-site/uploads/2021/12/WorldInequalityReport2022_Full_Report.pdf

PART 1

Transformations

1

CLIMATE TRANSFORMATION AND SOCIAL TRANSFORMATION

'Every politically significant revolution is anticipated by a transformation of the intellectual landscape'.

(Tony Judt[1])

'Each figure is a population and the stones – the trees, the muscled mountains are the world – but not the world apart from man – the world and man – the one inseparable unit, man and his environment'.

(John Steinbeck[2])

Introduction

Climate transformation and social transformation are intimately connected. The first cannot be properly addressed without understanding and promoting the second. The world today needs a new social transformation in order to manage the costs of climate change fairly and to guarantee democracy and social justice as the impacts of global heating sweep over us.

Much has been written in recent years about the threat to democracy from populist and crypto-fascist movements. I argue that the threat to democracy comes from within. The weakness of the state in responding effectively to inequality in societies has led large numbers of voters in democracies to turn to populist charlatans who seem to offer an alternative. Even an expert on foreign relations from a conservative US think tank argues that rising populism is fuelled by declining living standards and dysfunctional institutions.[3] But populists have no answer to climate change.

DOI: 10.4324/9781003382133-2

Nature is transforming under our feet and over our heads. We talk about *climate* change, but what is changing is the entire biosphere, the air above us, the land, the depths of the oceans, and all the creatures these environments support, including ourselves.

It is happening to the Earth's environment without which human societies cannot survive. It is also happening because human societies have been transformed as a result of one technological discovery, the use of fossil fuels to generate energy. That great discovery is now threatening to kill us.

Social scientists Joseph Camilleri and Jim Falk in 2009 ranked climate change as 'the defining issue of our time'. They told us that the world's governance capacities 'had to be reshaped across both time and space'. Spatially, decision-making processes and institutions have to extend their authority and legitimacy across the entire planet. Temporally, the ambit of government has to take account of the possibilities of the present, 'taking due account of current interests as well as past practices and traditions' (with varying contexts across the nations) but always acting to avoid the long-term consequences of climate change.[4]

The economist Thomas Piketty argues, 'What is certain is that the development of a sustainable climate policy will require new norms of environmental and fiscal justice that the majority can accept, which is fundamentally not the case today'.[5] Societies geared to manage climate change with fiscal justice and greater equality will also be more contented societies measured against well-being norms. Professors Richard Wilkinson and Kate Pickett demonstrated in their survey of data from 25 nations and 50 US states that 'equality is better for everyone'.[6]

The question we now face is what social transformation is needed to enable *all of us* to survive and *all of us* to prosper under the transformation of Nature that our human societies have generated?

Global Heating

The last time the Earth experienced atmospheric CO_2 concentration similar to the present was during the Pliocene epoch between five and three million years ago when the concentration ranged from 360 to 400 parts per million. The concentration of carbon dioxide in the atmosphere in 2022 was 417.2 parts per million – and growing.[7] In 2023, the global sea surface temperature jumped, hitting a new record high, the cause unknown.[8]

The United States Environmental Protection Agency graph of the change over time shows a regular upward curve of global atmospheric carbon concentration from 1950 to 2020.[9] The slope of the curve is not flattening as time goes by but becoming steeper. Carbon dioxide emissions are accompanied by gases such as methane and nitrous oxide and even water vapour which turbocharge global heating. As heating progresses natural processes break out, further heating the atmosphere.

The draft report of the Glasgow climate change summit (COP 22), *Impacts, Adaptation and Vulnerability, IPCC Summary for Policy Makers,* concludes with these words:

> The cumulative scientific evidence is unequivocal: climate change is a threat to human well-being and planetary health. Any further delay in concerted antici- patory global action on adaptation and mitigation will miss a brief and rapidly closing window of opportunity to secure a liveable and sustainable future for all.[10]

Overheating kills people. 'Over 70,000 excess deaths occurred in Europe during the summer of 2003' and 61,672 in 2022.[11] The year 2023, under a powerful El Niño, saw the hottest week (globally) ever recorded. Heat records were broken all over the Northern hemisphere, with raging bushfires from Canada to Greece. Polar ice is melting. An ice core extracted by the US army in the 1960s (rediscov- ered in 2023) indicated that Greenland had been free of ice 400,000 years ago with average temperatures similar to those of today.[12]

The Intergovernmental Panel on Climate Change (IPCC) now uses the term 'transformation' in its advice to policy makers. The transformation that is required, the IPCC Report says, 'changes the fundamental attributes of a social-ecological system in anticipation of climate change and its impacts'.[13] Even if all the policies now in place to limit global heating were fully implemented, the world would still warm by 3.2 degrees during this century. That means we are heading towards a Pliocene world by the end of the century.

The Earth in the Pliocene was in some ways recognisable. The hominid species (more than one), walking on two legs, were just beginning to emerge. But climati- cally the world was very different. Stunted shrubs survived the Antarctic winters, conifers grew in northern Greenland, grassy woodlands of larch and birch sup- ported a mammal population of bears, shrews, and wolverines in Arctic Canada. Deciduous forests became abundant. Coniferous forests grew where frozen tundra dominates today. Tropical forests remained only in a band around the equator. Grasslands spread on all continents except Antarctica. Some regions even cooled. Cloud cover over East Africa reflecting sunlight back into space cooled the region by two and three degrees.[14]

This may sound rather appealing, in part because the hotter climate was accompanied by increased rainfall and cloud cover in some areas. But the absence of continental glaciers in the Northern hemisphere contributed to a mean sea level 25 metres higher than today. Heating will not stop at Pliocene levels. If greenhouse gas concentrations are not stabilised, a three-degree rise in temperature will reverse the carbon cycle. Instead of absorbing carbon dioxide, vegetation and soils will release it back into the atmosphere. Methane which is today stored under the oceans and in the frozen tundra will become free to bub- ble up into the air.

Because of feedback effects which *will* start to occur once we exceed even 1.5 degrees of heating, the path to the neo-Pliocene and beyond will be self-sustaining. In the absence of radical intervention by humanity, every decade will take us closer, and every decade will contain more and more frequent catastrophic events: droughts, floods, overheated cities, air pollution, cyclones and tornados, firestorms over enormous areas, and rising sea levels threatening small islands and coastal cities.

Mark Lynas concludes his discussion of three degrees of warming with these words: 'As social collapse accelerates, new political philosophies may emerge, philosophies that lay blame where it truly belongs – on the rich countries that lit the fire that has now begun to consume the world'.[15]

In this book, we are going to examine the emergence of new philosophies, the *social* transformation necessary for nations to adapt to climate change. We will also see that it is not so much a matter of rich *countries* as of rich *people*, and not a matter of laying blame for the past but of enabling fairness in apportioning costs and encouraging a change of behaviour for the future.

Historical changes in technology present immense challenges for societies to adapt to new conditions in which their economies function. But successful adaptation is accompanied by ideological change in which injustice and inequality in society come to be debated, defined, and redefined. Modern forms of democracy have given the public a voice, and a choice over how much inequality can be tolerated in conditions of environmental and economic crisis.

Transition or Transformation?

Over the last 50 years, most observers of climate change have been hoping for a transition to a low or zero carbon-emitting global economy. The technology has existed for years to make the transition, seemingly without major disruption to human lifestyles. It seemed entirely feasible that the transition could happen.

There has been a transition. But it is not sufficient to stop climate change. While individuals and firms can help to reduce emissions and shape public opinion in favour of strong action on the climate emergency, the necessary social transformation requires action by governments, nation-states, and international regimes. Transformation must aim to change the rules, the institutions of society, to ensure that the costs of global heating are shared fairly. States are the only entities with the legitimacy to make the rules for society and to enforce those rules.

There is a growing literature of 'sustainability transformations'. Linnér and Wibeck write of 'the complex entanglement between past and present that is often just below the surface of everyday relations'.[16] Edmondson presents a collection of essays on 'Sustainability transformations, social transitions and environmental accountabilities'.[17] The main focus of this literature is human–environmental relations. My focus in the present book is the transformation of society and what we can learn from historical examples, an approach I describe as 'critical history'.

What then is the *transformation* of society? Superficially, we can think of 'transformation' as a more or less peaceful, more or less deliberate, and more or less profound institutional change. There is an important distinction between transition and transformation. A *transition* can be said to mean an economic or technological change without altering the ideological settings of society.[18] A *transformation* means economic and social change in which the ideological settings are changed – a change of political ethos, a change in what is known to be true and what is known to be right.

I make a distinction between transformation and revolution. The latter is a profound institutional change achieved by violent means, often resulting in civil war. There is no doubt that both the French and Russian revolutions achieved institutional change well beyond the bounds of the two nations themselves, but at great social cost – and in both cases, eventual backsliding.

By transformation, I mean fundamental social change without violence. Today, we need to plan for a peaceful transformation to occur which will meet the challenge of climate change. If a new transformation is required, how will it occur, what will it look like, and will it endure?

Social Transformation

To address the transformation of society, I start with a simple threefold conceptual template: that social transformation occurs as a result of a building sense of crisis within society and polity; that transformation entails a change in ideology; and that transformation requires political movements and champions.

But there is a deeper meaning of transformation analysed by Karl Polanyi in his seminal work, *The Great Transformation*. This transformation is from a world of feudal societies whose cohesion is organised by traditional relationships and power structures to a world of societies organised primarily by market relationships. This transformation is global and ongoing.

In the first part of the book, I explore social transformation by examining the historical record, starting with *The Great Transformation*. I then examine three subsequent social transformations (embedded as they were within the broad development of capitalism) that occurred in the twentieth century: the transformation to social democracy in Britain and Europe during and after the Second World War, the transformation from social democracy to neoliberalism in the 1980s, and the transformation of communism to capitalism in Russia and its client states.

All these transformations were felt like social tsunamis worldwide. The first, beginning with Lloyd George's 'People's Budget' and culminating in the welfare state, swept away the cobwebs of aristocratic power and gave effect to the people's democracy. The second started a fightback by the owners of wealth that disabled effective democracy. The third, under the influence of the second, sent Russia into convulsions of humiliation that led ultimately to fascism and the end of peace in Europe.

Neither pure Marxism nor pure liberalism works. A market economy will produce what consumers want *and can pay for*. This is called effective demand. Inequality of income and wealth shapes effective demand. If income and wealth are concentrated at the top of the scale (top 1 per cent say), the economy will be skewed to what the top of the scale wants. The world's wealth is flowing into the production of the most expensive luxury goods while the needs of the majority are not met – the world's richest person is not Musk or Gates but Bernard Arnault, CEO of the corporation LVMH valued at US$708 billion. LVMH owns luxury brands Dior, Louis Vuitton, Moët & Chandon champagne, and Hennessy cognac.

The followers of Marx believed that the way to correct that problem was to put the working class in charge of the state – through revolution – and nationalise the means of production and exchange, substituting state economic planning for the market. It was tried and it failed.

While there existed no effective democracy, the Marxian solution seemed the only way to ensure that what the large majority wanted and needed was produced. But once effective majoritarian democracy was established, there was another solution: *legislate* to reduce inequality and allow the market to respond. This second solution turned out to be extremely effective, while total state ownership and a politically directed economy ultimately failed. The freedoms and utility of markets could function, while the state ensured a fair distribution of income and wealth.

In the second part of the book, I turn to today's crises: of global heating, of democracy, and of inequality and poverty. I then interrogate the leading example of a worked-out social democratic ideology dealing with all three interconnected crises. I draw heavily on the work of Thomas Piketty and his colleagues at the World Inequality Lab based at the University of California Berkeley and the Paris School of Economics. Finally, and by way of a conclusion, I speculate on how a new social transformation might unfold where, when, and with what actors and champions.

Notes

1 Judt, T. (2005) p. 535.
2 Written 'during the composition of *To a God Unknown* in 1930'. Shillinglaw, S., (2009) 'Introduction' to Steinbeck's novel *Cannery Row,* Penguin Books.
3 Mounk, Y. (2021).
4 Camilleri, J. and Falk, J., (2009).
5 Piketty, T. (2020) p. 1007.
6 Wilkinson, R. and Pickett, K. (2010) *The Spirit Level, Why Equality is Better for Everyone*, London: Penguin Books. The authors tested the levels of income inequality against data (from 23 rich countries and 50 US states) on physical and mental health, life expectancy, infant mortality, social cohesion and trust, educational performance and welfare, homicides and imprisonment, and the connection between well-being and sustainability. In many countries today 'well-being' is being measured as an index of economic performance https://www.oecd.org/wise/measuring-well-being-and-progress.htm (accessed 24/07/20123).

7 National Oceanic and Atmospheric Administration (USA), NOAA Research News, https://research.noaa.gov/2022/11/15/no-sign-of-significant-decrease-in-global-co2 -emissions/#:~:text=The%20publication%2C%20produced%20by%20an,percent %20above%20pre%2Dindustrial%20levels (accessed 03/07/2023).
8 https://www.bbc.com/news/science-environment-65339934 (accessed 27/04/2023).
9 https://www.climate.gov/news-features/understanding-climate/climate-change -atmospheric-carbon-dioxide (accessed 27/04/2023).
10 United Nations (2022).
11 Ballester, J., Quijal-Zamorano, M., Méndez Turrubiates, R.F. et al. (2023), 'Heat-related mortality in Europe during the summer of 2022'. *Nature Medicine* 29, 1857–1866 (https://doi.org/10.1038/s41591-023-02419-z).
12 Bierman, P. (2023) 'When Greenland was Green: Ancient soil from beneath a mile of ice offers warnings for the future', *The Conversation*, 21/07/2023, https://theconversa-tion.com/when-greenland-was-green-ancient-soil-from-beneath-a-mile-of-ice-offers -warnings-for-the-future-209018 (accessed 22/07/2023).
13 United Nations (2022), p. 5.
14 Lynas, M. (2008) pp. 131, 132.
15 Ibid., p. 181.
16 Linnér, B-O and Wibeck, V. (2019) Preface.
17 Edmondson, E. (Ed.) (2023).
18 As in, for example, Bridge, G, Bouzarovski, S, Bradshaw, M., and Eyre, N (2013). The authors show how historical energy transitions 'have been associated with broad social change, such as industrialisation, urbanisation, and the growth of the consumer soci-ety' (p. 333), but they do not address the causes (other than technological) of such social change.

References

Bridge, G., Bouzarovski, S., Bradshaw, M., and Eyre, N. (2013) 'Geographies of energy transition: Space, place and the low-carbon economy', *Energy Policy*, 53, 331–340.

Camilleri, J. and Falk, J. (2009) *Worlds in Transition, Evolving Government Across a Stressed Planet*, Cheltenham: Edward Elgar.

Edmondson, E. ed. (2023) *Sustainability Transformations, Social Transitions and Environmental Accountabilities*, London and New York: Palgrave-Macmillan.

Judt, T. (2005) *Post War, A History of Europe Since 1945*, London: Penguin Books.

Linnér, B.-O. and Wibeck, V. (2019) *Sustainability Transformations, Agents and Drivers Across Societies*, Cambridge: Cambridge University Press.

Lynas, M. (2008) *Six Degrees, Our Future on a Hotter Planet*, Washington, DC: National Geographic.

Mounk Y. (2021) 'After Trump, Is American Democracy Doomed by Populism' Council on Foreign Relations, *In Brief*. https://www.cfr.org/in-brief/after-trump-american -democracy-doomed-populism (accessed 08/05/2023).

Piketty, T. (2020) *Capital and Ideology*, Cambridge: Bellknap Press of Harvard University.

United Nations. (2022) *IPCC WGII Sixth Assessment Report*, Summary for Policy Makers, SPM.D.5.3.

2

THE GREAT TRANSFORMATION

Introduction

We know that profound change in the way human society is organised has occurred during the last two or three centuries – an extraordinarily short period in the history of humanity. Some social scientists call this an epochal change from a state in which humanity was subject to Nature (the biosphere), to a state in which Nature is subject to the activities of humanity: the Anthropocene epoch.

We need explanations of social transformation: not only of the different *states* of society before and after transformation, but of the *process* of transformation: how and why transformations have occurred.

In the twentieth century, the discursive space of transformation was dominated by the work of Karl Marx and Friedrich Engels, and of the scholars and activists interpreting their work for the present, building on it or, in various ways, reacting to it.[1] They have contributed much to our understanding of social transformation.

The work of Karl Polanyi offers an alternative to Marx, at least as profound and at least as historically grounded. It also has the advantage of knowing what occurred in human society during the hundred years since Marx was writing, as well as taking account of the history of capitalism in the two centuries or more before Marx. Moreover, Polanyi's work, like that of Marx and Engels, is grounded in the values of socialism. Most importantly, it records the transformation of democracy.

Marxism and its variants are unable or unwilling to offer a vision of a democratic, pluralist, socialist society that protects individual freedoms, and certainly not one that accommodates capitalist markets. In the Marxist perspective, capitalism must be abolished. Democracy represented bourgeois values supportive of a social system created by and for the owners of wealth. Polanyi, on the other hand,

DOI: 10.4324/9781003382133-3

points us to actual democratic socialism not only co-existing with market societies but making market societies possible.

Polanyi interpreted the development of capitalism in terms of a dynamic and shifting relationship between, on the one hand, the creation, expansion, and justification of a self-regulating market, and on the other, the protection of humanity and nature from the effects of such a market. The two 'movements' had to occur together, or else either industrial development with all its material benefits to humanity would be destroyed or humanity and nature would be destroyed. Thus, capitalism needed socialism in order to prosper, and socialism grew out of capitalism.

Polanyi's analysis in *The Great Transformation* ends in the glow of the impending reality of socialism within a market society – indeed in the originator of market society, the United Kingdom. As we know that glow was dimmed, social democracy reversed from the 1980s – though not without some continuing social protections.

Polanyi explains at the start of his book: 'Our thesis is that the idea of a self-adjusting market implied a stark utopia. Such an institution could not exist for any length of time without annihilating the human and natural substance of society; it would have physically destroyed man and his surroundings into a wilderness'.[2] Though meticulously situated in the history of Britain's transformation from a patriarchal and agrarian society into a modern industrial economy, Polanyi states that his is not a historical work but rather an explanation of the trend of events *in terms of human institutions.*

To put the matter in more contemporary terms, 'the state cannot be understood in isolation from the contingencies of contemporary political, cultural, and economic life'. This understanding must bring into focus 'the relationship between the state and civil society on the one hand, and between the state and the market on the other'.[3]

The purpose of this chapter is not to provide an exegesis of Polanyi's text but to discuss the ideas that resonate with the social circumstances of today. In what follows, we will depart from Polanyi's order and group the discussion under four main headings: the connection between the transformation of production and the ideology both of the self-regulating market and of nascent socialism, the growth and necessity of the protective movement, and the relationship between the market dynamic and 'Man and Nature'. Finally, we consider the international order, its fragility, and the alternatives of authoritarian fascism and democratic socialism.

Transformation

The industrial revolution began in Britain with the invention of the coal-powered steam engine. That revolution in the application of science and technology to production continues its sweep across the globe today, with ever new technologies, ever new sites of invention and development, and ever new locations of production. The inventions of industrial agriculture, the internet, digital communication, artificial intelligence, robotics, medicines, vaccines, and instruments of warfare have generated waves of techno-revolution breaking over society.

The technical revolution inevitably led to a market-based society. Polanyi writes: 'We do not intend to assert that the machine caused that which happened, but we insist that once elaborate machines and plant were used for production in a commercial society, the idea of a self-regulating market system was bound to take shape'.[4] The revolution was, as Polanyi writes, 'an almost miraculous improvement in the tools of production'. But its effects on society and environment were devastating. Even 40 years before the full impact of industrialisation was unleashed, there began to develop institutions of social protection.

During the eighteenth century, both land and money were mobilised. In Britain, a wave of enclosures ensured that land could be developed anywhere in the nation, investment directed anywhere. The House of Lords and their allies in the House of Commons enacted 'ferocious' laws to assert and protect the sacred right to property. 'Their purpose was to put hedges around fields and put an end to the right of poor peasants to use communal land for crops and pasturage'.[5] These enclosure laws were paralleled by the so-called Black Act of 1723 'which stipulated the death penalty for anyone caught pilfering wood or poaching game on land they did not own'.[6]

A new class of employers was emerging, but labour was left immobile. Before 1795 in Britain, the formation of a national labour market was 'inhibited by strict legal restrictions on physical mobility, since the labourer was practically bound to his parish'[7] (the smallest unit of local government). If jobs were not available where workers lived, and if they could not move to where there were jobs, they would starve.

To deal with the pernicious effects of rural poverty, the Berkshire County Court of Quarter Sessions held a meeting at the village of Speenhamland near Newbury (a market town west of London), the outcome of which was a new allowance system that subsidised wages from local property taxes – the 'rates' – linked to the price of bread. The system specified a minimum wage that would prevent starvation, or in the absence of such a wage, an allowance of money to support the worker and his family. It was called 'outdoor relief' as opposed to the 'indoor' relief provided by the poorhouse. This system was intended as an emergency measure of social protection. It became adopted and established over most of rural England and even in some industrial towns.

The Speenhamland system did prevent starvation and allowed for a gradual adjustment of society to a free labour market. But it also had severe unintended consequences. First, it created a situation where workers had no incentive to work nor employers to increase productivity. It allowed employers to pay no more than the minimum wage and therefore ended up in effect as a subsidy to employers. It created a new class of 'paupers' dependent on 'the rates' for survival.

Polanyi describes the situation: 'Once the intensity of labour, the care and efficiency with which it was performed, dropped below a definite level, it became indistinguishable from "boondoggling" or the semblance of work maintained for the sake of appearances'. Yet Speenhamland was widely popular: 'parents were free of the care of their children, and children were no more dependent upon parents; employers could reduce wages at will and labourers were safe from hunger whether they were busy or slack'.[8]

What was ruined was not life as such, but the dignity of labour, the self-esteem of workers, and their sense of a place in the culture and the life of a community. In conjunction with the Anti-Combination Laws (banning the formation of labour unions), which were not revoked for another quarter century, 'Speenhamland led to the ironic result that the financially implanted "right to live" eventually ruined the people whom it was ostensibly designed to serve'.[9]

The devastating failure of Speenhamland should put us in mind of a contemporary well-intentioned failure – even if not so devastating. Without a plentiful supply of affordable housing, financial assistance for first home buyers, either from governments or drawn from compulsory personal retirement funds, invariably ends up increasing the price of housing, thus benefiting existing homeowners and making it harder for new home buyers to become homeowners, for example, the American *Downpayment Toward Equity Act* of 2021, the *Help to Buy: Equity Loan* in the UK, and various *First Home Buyer Assistance Schemes* in Australia.

The Speenhamland system was dismantled under the Poor Law Reform Act of 1834, which cancelled the 'right to live' and freed up the labour market for work in the 'satanic' mills of industry. With that Act and all that followed from it, capitalism arrived suddenly and 'unannounced'. With it came the working class, poverty in a society which was creating untold wealth, the very discovery of 'society' itself as a national phenomenon, and the 'fictitious' commodities: labour, land, and money.

Pauperism (living 'on the rates') was abolished, but the poor remained, a new industrial class.[10] The condition of the working class of England in the decade that followed was appalling, well known, and well documented.[11] Polanyi reminds us:

For some seventy years, scholars and Royal Commissions alike had denounced the horrors of the Industrial Revolution, and a galaxy of poets, thinkers, and writers had branded its cruelties. It was deemed an established fact that the masses were being sweated and starved by the callous exploiters of their helplessness; that enclosures had deprived country folk of their homes and plots, and thrown them on the labour market created by Poor Law Reform and that the authenticated tragedies of the small children who were sometimes worked to death in mines and factories offered ghastly proof of the destitution of the masses.[12]

But as he explains:

> A blind faith in spontaneous progress had taken hold of people's minds, and with the fanaticism of sectarians the most enlightened pressed forward for boundless and unregulated change in society. The effects on the lives of the people were awful beyond description.[13]

Market Ideology

What ideological construction was devised to explain and justify this situation? What, in fact, explains 'ideology'? Systems of reasoning, or ideology, are required to shape institutions. Institutions are required to shape society. In the nineteenth century, the belief that human society should be subordinated to self-regulating markets 'became the organizing principle for the world economy'.[14]

However, humans in all stations of society and all classes have the capacity for empathy. Insofar as we are human (and not suffering from psychopathy), we are aware of the suffering of others and in some way feel the pain ourselves. The capacity for empathy amongst the powerful as well as protest from its victims triggered a reaction to the destructive effects of the self-regulating market on humanity, giving rise to the birth of socialism.

At its beginning, the industrial revolution and its development of machine production needed the transformation of the natural and human substance of society into commodities in a commercial world. The commodity fiction supplied a vital organising principle of a market society, but, 'To allow the market mechanism to be sole director of the fate of human beings and their natural environment indeed, even of the amount and use of purchasing power, would result in the demolition of society'.[15] For Polanyi, that contradiction drove, more or less simultaneously, two opposed strands of ideology, on the one hand the liberal doctrine of the self-regulated market and on the other the socialist doctrine of protection of humanity and nature.

At the end of the eighteenth century, British classical economists observed that economic growth corresponded with the rise of poverty. As Polanyi describes the situation, that moment also corresponded with a period of slump in trade following the Seven Years War. However, poverty came to be regarded as instrumental to the creation of wealth. The liberal doctrine of the self-regulating market emerged from a collision between hope and despair. A vision of hope was distilled out of the 'nightmare of population and wage laws and was embodied in a concept of progress so inspiring that it appeared to justify the vast and painful dislocations to come'.[16]

The key strands of the web of liberal ideology regarding humanity, as Polanyi tells it, came together to form a justification of poverty and wealth, in short of inequality. Joseph Townsend argued that laws that determined that workers shall never hunger are misconceived. Laws may certainly compel workers to work but legal constraint is 'attended with much trouble, violence and noise; creates ill will and never can be productive of good and acceptable service'.[17]

Hunger, on the other hand, is peaceful, silent, and exerts unremitting pressure. Slaves must be compelled to work, but free men should be left to their own judgement under the constraint of hunger. From this novel point of view, Polanyi writes, 'a free society could be regarded as consisting of two races: property owners and labourers. The number of the latter was limited by the amount of food; as long as property was safe, hunger would drive them to work'.[18] Beyond noting Townsend's comparison between slave societies and free societies, Polanyi has little to say about the slave trade on which much of Britain's middle-class prosperity was built. Piketty on the other hand devotes a long chapter to the 'extreme inequality' of slave societies.[19]

For Edmund Burke, the abolition of poor law relief for the able-bodied unemployed without minimum wages being set would enable labour to be treated as it really was, a commodity which would find the level of its wages in the market.[20]

For Jeremy Bentham, poverty was 'nature surviving in society'. In the highest stage of social prosperity, he thought, the mass of citizens would most probably live with few other resources beyond their daily labour and consequently would always be near to indigence.[21] The task of government was to increase want in order to make the physical sanction of hunger effective. Yet 'the community' (meaning those with power) would not completely disregard the fate of the indigent. Relief would be provided within his 'industry houses' described by Polanyi as 'a nightmare of minute utilitarian administration enforced by all the chicanery of scientific management'.[22]

The dismal view that in order to create prosperity humanity would need to be subjected to the 'law of the jungle' proved wrong. Adam Smith's view was that universal plenty could not help percolating down to the people: 'it was impossible that society should get wealthier and wealthier and the people poorer and poorer'.[23] The 'naturalistic' iron law of wages – that wages of workers were fated by nature to remain at subsistence level – was 'absurd from the point of view of any consistent theory of prices and incomes under capitalism'. But it took another century after publication of *The Wealth of Nations* for it to be clearly realised, as Smith continues, 'that under a market system the factors of production shared in the product, and as produce increased, their absolute share was bound to rise'.[24]

The truth of this statement is borne out by the facts – though only when capitalism is tempered by social protection and redistribution. Piketty in *A Brief History of Equality* insists that 'human progress exists' as is demonstrated by the dramatic rise in worldwide life expectancy, literacy and average income per month and per person since the beginning of the nineteenth century.[25] Yet, had Polanyi been writing after the reactionary neoliberal transformation of the 1980s, he might have observed that the ghost of Townsend and Bentham lingers in the current attitude to poverty and the treatment of the unemployed in Britain, the USA, and Australia.

The Emergence of Socialism

Polanyi traces the emergence of socialism within the market society being created in Britain to the Quaker philosopher John Bellers in 1696 and much later to Robert Owen from 1819. In the late seventeenth century, Quaker 'Meetings of Sufferings' were using statistics to support their religious beliefs. Bellers argued that 'the labour of the poor being the mines of the rich', the poor should be able to support themselves by exploiting those riches (their own labour) for their own benefit.[26]

All that was needed was to organise workers into a 'college' or corporation where they could pool their labour. Polanyi remarks that the idea of 'colleges of labour' was at the heart of all later socialist thought on the subject of poverty, whether it took the form of Owen's *Villages of Union*, Fourier's *Phalanstères*, Proudhon's *Banks of Exchange*, Louis Blanc's *Ateliers Nationaux*, Lassalle's *Nationale Werkstätten*, or, for that matter, Stalin's *Five Year Plans*.[27]

Labour, and not money, was to be the standard of value. Subsistence and payment according to results were to be combined. Any surplus in value was to return to the community. The concept of economic self-sufficiency of the labouring class inspired the early Trade Union movement: as a general association of all trades, crafts, and arts in One Big Union in the body of society. Such an idea was, in embryo, the source of syndicalism, socialism, anarchism, and even capitalism in their plans for the poor. Robert Owen, however, stood out as a man of vision who had close first-hand experience of industry.

The son of a Welsh saddler and ironmonger, at the age of 19 Owen borrowed 100 £ to set up a business manufacturing spinning machinery. He sold his share of the business in 1789 and went to manage a mill with 500 workers. In the course of his work at the factory he met many businessmen in the textile industry including David Dale, the owner of the Chorton Twist Company in New Lanark, Scotland. Supported by several other businessmen, Owen bought Dale's four textile factories in New Lanark and set out to make New Lanark a model for a village reflecting his philanthropic ideas.[28]

Owen had experienced, as a manager, the 'ordeal' of the factory. Of all the theorists of liberalism, he alone had intimate personal knowledge of industry. He was aware both of the promise of the machine for material improvement and the important role of the state in averting harm to the community. Polanyi tells us that neither the political mechanism of the state nor the technological apparatus of the machine hid from him *the* phenomenon of society. The fulcrum of his thought, Polanyi asserts, lay in his criticism of Christianity, which he accused of individualisation, thus denying the social origin of human motives.

From his observation of factory life, Owen developed the belief that a person's character is formed by the effects of their working environment. With the creation of his philanthropic venture, he is considered a founder of the idea of town planning as a social institution. He opposed harsh treatment of factory workers. He insisted that education was critical to developing good and humane people, so he built nursery and infant schools and provided play spaces for children. Older children worked in the factory but attended secondary school in New Lanark.

Polanyi quotes Owen directly (in 1819):

> The general diffusion of manufactures throughout a country generates a new character in its inhabitants; and as this character is formed upon a principle quite unfavourable to individual or general happiness, it will produce the most lamentable evils, unless its tendency be counteracted by legislative interference and direction..[29]

Polanyi concludes that,

> Owen knew that the industrial revolution was causing a social dislocation of stupendous proportions, and the problem of poverty was merely the economic aspect of this event. Owen justly pronounced that unless legislative interference and direction counteracted these devastating forces, great and permanent evils would follow.[30]

Robert Owen's philosophy inspired a widespread proto-socialist movement seeking to found industrial communities based on the principles established for New Lanark. Owen himself went to North America in 1824 to spread his ideas for co-operatives, labour exchanges, and experimental communities. Owenite communities sprang up in England, Ireland, America, and Canada.

Polanyi writes, 'The Owenite Movement originally was neither political nor working class. It represented the cravings of the common people, smitten by the coming of the factory, to discover a form of existence which would make man master of the machine'.[31]

Unlike Owenism, the Chartist Movement that erupted in the ten years from 1838 'appealed to a set of impulses' very different from those of Owenism.[32] The Chartists attempted to put pressure on the middle classes (which had secured the vote thanks to the Reform Movement) to accept effective popular suffrage. The Chartists demanded nothing less than participation in parliamentary government by working people through universal male suffrage and reform of the electoral institutions.[33] Chartism was a movement for *democracy*, not for the abolition of *capitalism*.

The Movement needs to be understood within the history of its times, Polanyi argues:

> The years 1789 and 1830 made revolution a regular institution in Europe; in 1848 the date of the Paris rising was actually forecast with a precision more usual in regard to the opening of a fair than to a social upheaval, and 'follow up' revolutions broke out promptly in Berlin, Vienna, Budapest and some towns in Italy.

Polanyi records that in 1848, 'In London ... there was high tension, for everybody including the Chartists themselves expected violent action to compel Parliament to grant the vote to the people'. Yet the Charter was uncompromisingly rejected by the government and crushed with the mobilisation of an unprecedented force of hundreds of thousands of armed citizens recruited as 'special constables' for the maintenance of law and order.

The Chartists dispersed peacefully. The Paris revolution of 1848 'came too late to carry a popular movement in England to victory'. Besides, the spirit of revolt roused by the Poor Law Reform Act was already waning: 'the wave of rising trade was boosting employment, and capitalism began to deliver the goods'.[34] Democratic reform had to wait another 30 years. Polanyi points out that only when the working class and the unions had accepted the principles of a capitalist economy did the middle classes concede the vote to the better situated workers.[35]

The history of liberal and democratic socialist ideology and their supportive interests now needs to be set within a wider discussion about the movement for social protection.

Transformation and Social Protection

It is Polanyi's thesis that the movement for social protection includes but cannot be reduced to the action of material class interests. The double movement grew within the transformation of society wrought by the industrial revolution. It is worth quoting Polanyi at length here.

> It [the double movement] can be personified as the action of two organizing principles in society, each of them setting itself specific institutional aims, having the support of definite social forces and using its own distinctive methods. The one was the principle of economic liberalism, aiming at the establishment of a self-regulating market, relying on the support of the trading classes, and using laissez-faire and free trade as its methods; the other was the principle of social protection aiming at conservation of man and nature as well as productive organization, relying on the varying support of those most affected by the deleterious action of the market – primarily, but nor exclusively, the working and the landed classes – and using protective legislation, restrictive associations, and other instruments of intervention as its methods.[36]

Polanyi argues that once we rid ourselves of the 'obsession' that only sectional monetary interests and never general interests can become politically effective, the reason for the broad comprehensiveness of the movement for social protection becomes clear. While monetary interests are 'necessarily voiced solely by the persons to whom they pertain', there are other interests with a wider constituency. They belong to individuals as neighbours, professional persons, consumers, pedestrians, commuters, sportsmen, hikers, gardeners, patients, mothers, or lovers. They find representation by almost any kind of territorial or functional association: churches, townships, fraternal lodges, clubs, trade unions, or political parties based on broad principles.[37]

The concept of a pluralist socialism based on such a perception was eclipsed by the monist variant created by Marx and implemented by Lenin in Russia. The influence of the communist variant spread throughout the world and is felt to this day. But pluralist socialism can be traced through the work of American political theorists Charles Lindblom and Robert Dahl in their later work.

Based on their research into American politics, they argued that the owners and managers of 'big businesses' possessed a powerful hold over democracy through their control over investment. The logical conclusion that Dahl in particular draws is that the management of large companies should be subject to democratic accountability in the same way as states. Such arrangements have been implemented in Scandinavia and Germany with workers' representatives sitting on the boards of large companies. As we will see in Chapter 10, Thomas Piketty continues that theme of pluralist socialism in *Capital and Ideology* (2021).

Polanyi's is as clear a statement of the fundamentally pluralist nature of human interests as one can hope to find. To be clear, different classes can rise to take leading roles in social development at different times: 'Frequently, at a historical juncture new classes have been called into being simply by the demands of the time'.[38]

Today, if we seek to explain social and political events, we need to be clear about the nature of social forces – classes or otherwise – existing today, and not to force our analysis into the shape of such forces existing in the time of Marx, or even of Polanyi himself.

Moreover, class interests offer only a limited explanation of long-run movements in society. The birth and death of classes, their aims and the degree to which they succeed in attaining them, their alliances, and antagonisms cannot be understood apart from the interests of society as a whole 'given its situation as a whole'. In a passage which resonates with humanity's situation today – dealing with pestilence, the opportunities and perils of the digital economy, international warfare and global heating – Polanyi writes,

> Now, this situation is created, as a rule, by external causes, such as a change in climate, or the yield of crops, a new foe, a new weapon used by an old foe, the emergence of new communal ends, or for that matter, the discovery of new methods for achieving the traditional ends.[39]

Polanyi concludes:

> An all too narrow conception of interest must in effect lead to a warped vision of social and political history, and no purely monetary definition of interests can leave room for that vital need for social protection, the representation of

which commonly falls to the persons in charge of the general interest of the community – under modern conditions, the governments of the day.[40]

Classes and Society

As the organising principle of the market spread through society, new alignments of class interests developed: on the side of social protection, the landed classes sought the solution to all the evils of industrial society in the past, while the industrial workers looked to a future 'commonwealth of labour'. On the side of material improvement through industry, a new class of entrepreneurs formed out of the remnants of older classes.

These developments resulted in new ideological claims. Liberal theory had first to explain the utter devastation of society in conditions of the surging growth of production and wealth, and second the countervailing presence of the widespread protective movement. Liberals relied upon the materialist philosophies of Ricardo and Bentham. According to the accepted yardsticks of economic welfare, the social inferno of early industrial capitalism never existed. The working classes were never exploited and were the economic winners from emergent capitalism. So, what could explain the movement for social protection?

The appearance of the Chartist Movement seemed to support what Polanyi calls 'the collectivist conspiracy' of Liberal ideology or, in crude Marxist ideology, the 'class theory of social development'. Liberals and Marxists alike deduced the protectionist movement from the force of material sectional interests. It is perhaps no surprise that, once Marxism was dead and buried after 1989, Russian economists nurtured under Marxist theory (as we will see in Chapter 5) had no great difficulty in accepting liberal economics based on the harsh doctrine of utilitarian principles.

Polanyi sets out to show precisely and in detail that 'the testimony of the facts contradicts the liberal thesis decisively. The anti-liberal conspiracy is pure invention'. The great variety of forms, he argues, 'in which the "collectivist" countermovement appeared was not due to any preference for socialism or nationalism on the part of concerted interests, but exclusively to the broad range of the vital social interests affected by the expanding market mechanism'.[41]

Polanyi's historical 'testimony of the facts' is set out in forensic detail in two key chapters.[42] He argues:

1. There is an amazing diversity of the matters on which protective action was taken; this alone would exclude the possibility of concerted action.
2. The change from liberal to collectivist solutions happened sometimes overnight and without any consciousness on the part of those engaged in the process of legislative rumination.
3. There is a remarkable coincidence of similar legislative development in different countries 'of a widely dissimilar political and ideological orientation'.[43]

For instance, Workers' Compensation was enacted in England (1880–1887), Germany (1879), Austria (1887), and France (1889). Factory inspection was introduced in England (1833), Prussia (1853), Austria (1883), and France (1874 and 1883).

4. Under the most varied slogans, with very different motivations, a multitude of parties and social strata put into effect almost exactly the same measures (of social protection) in a series of countries in respect of a large number of complicated subjects. Economic liberals themselves advocated restrictions on the freedom of contract and laissez-faire at various times.

In sum, Polanyi argues, 'The countermove against economic liberalism and laissez-faire possessed all the unmistakable characteristics of a spontaneous reaction. At innumerable disconnected points it set in without any traceable links between the interests directly affected or any ideological conformity between them'.[44] The change from laissez-faire to collectivism took place in different countries at definite stages of industrial development in a way that cannot be credited to changing moods or sundry interests.

Yet theorists of liberalism continued in their belief that labour was simply a commodity and that trade unions with their struggles for fair compensation and better working conditions disrupted the proper working of the labour market. The movement for social protection was in the view of Mises (and those of his utopian liberal co-believers), a disruption to the perfect working of the market driven by impatience, greed, and short-sightedness, 'but for which the market would have resolved its difficulties'.[45]

With Mises we can trace the market utopia through to the twentieth century and the work of Mises's pupil Friedrich von Hayek, and through him to the neoliberal ideological basis of the transformation initiated politically by Margaret Thatcher (see Chapter 4).

Today we have global free movement of capital, but movement of labour is for the most part constricted by rules relating to national boundaries. We have reactions in the form of Brexit and other nationalist anti-immigrant movements around the world. Such movements today are making the same mistake that the squirarchy and local judiciary made in nineteenth-century Britain with the Speenhamland system. In global capitalism, nation-states are constrained to prevent starvation with minimal welfare payments at the same time as suffering skills and labour shortages. But mobility of labour is deeply problematic as Polanyi argues below.

Let's turn now to Polanyi's critique of the relationship between market society and what Polanyi describes as 'fictional commodities' in a world interpreted by market ideology: humans ('man'), nature ('land'), and money.

The Market and Man[46]

Polanyi shows how the liberal utopia had to imagine labour as a commodity to be exchanged just like the things produced for sale. Labour is thus reified – becomes thing-like, which is in fact a monstrous denial of the humanity of working people, and thus of natural human rights, a concept Bentham denounced as nonsense on stilts. But Polanyi makes a further point that resonates today in issues of colonialism and its consequences.

Polanyi argues that it is not economic exploitation but the disintegration of the cultural environment of human society that is 'the cause of degradation'.[47] He posits that the Industrial Revolution in Britain had cataclysmic consequences for the culture of societies: 'an economic earthquake which transformed within less than half a century vast masses of the inhabitants of the English countryside from settled folk into shiftless migrants'.[48]

To be sure, both before and after the Industrial Revolution, the common 'folk' were exploited economically by the powerful, but only with the imposition of market society by the instruments of the state was their social condition, their culture, their sense of purpose, and their settled relationships with their fellows totally disrupted. 'The economic process', Polanyi argues,

> may, naturally, supply the vehicle of destruction, and almost invariably economic inferiority will make the weaker yield, but the immediate cause of his undoing is not for that reason economic; it lies in the lethal injury to the institutions in which his social existence is embedded.[49]

Polanyi draws a parallel between Britain in the aftermath of the Industrial Revolution and the circumstances of colonised societies in Africa: 'To the student of early capitalism the parallel is highly significant. The condition of some native tribes in modern Africa carries an unmistakable resemblance to that of the English labouring classes during the early years of the nineteenth century'.[50]

The clash of cultures across the world, from the Americas to Northern Europe, from Africa and East Asia to Australia and the Pacific Islands continues to carry reverberating economic and cultural consequences. In the interests of profit from food production or mining, it makes little difference whether the indigenous population works under the direct supervision of the colonist or only under some form of indirect compulsion, for in every and any case, Polanyi argues, the social and cultural system of indigenous life must be first shattered.

The example of Australia is instructive. The white settlers brought with them the institutions and culture of nineteenth-century Britain with all its assumptions about the sanctity of property and the institutions of government and law (although some institutions in the early days were honoured in the breach!). But Australia's indigenous peoples had over many thousands of years developed a settled culture with its own institutions, values, languages, national boundaries, mythologies, trading, and cultivation of the land. All of which were completely unrecognised by the colonists.

While the colonial adaptation of British institutions in Australia was in many ways democratically superior to those of England, in the course of colonisation, indigenous society was simply denied and wiped out with disastrous social consequences from which that society is only in recent years beginning to recover.

Polanyi concludes,

> To argue that social legislation, factory laws, unemployment insurance, and, above all trade unions have not interfered with the mobility of labour and the flexibility of wages, as is sometimes done, is to imply that those institutions have failed in their purpose, which was exactly that of interfering with the laws of supply and demand in respect to human labour, and removing it from the orbit of the market.[51]

The Market and Nature

We have also seen that Polanyi writes about the intimate connection between humans and 'land': 'Labour and land are no other than the human beings themselves of which every society consists and the natural surroundings in which it exists'.[52] First Nations people have made us aware of the powerful spiritual and emotional connection between indigenous peoples and the landscapes they traditionally occupy.

Despite the global mobility of goods and capital, in many respects human life remains obstinately 'fixed in place'. Political sovereignty is for that reason fundamentally territorial. The appalling conditions of nineteenth-century urbanisation could be gradually improved by the action of the protective movement and government at local and national levels. Protection could also be extended to,

> the conditions of safety and security attached to the integrity of the soil and its resources – such as the vigour and stamina of the population, the abundance of food supplies, the amount and character of defence materials, even the climate of the country which might suffer from the denudation of forests, from erosion and dust bowls, all of which, ultimately depend on the factor land, yet none of which respond to the supply-and-demand mechanism of the market.[53]

Polanyi's perspective is 'planetary' (inclusive of the environment) rather than 'global' (exclusively economic). 'With free trade new and tremendous hazards of planetary interdependence sprang into being'.[54] He understands the social effects of globalisation as an accomplished fact.

> The mobilization of the produce of the land was extended from the neighbouring countryside to tropical and subtropical regions – the industrial-agricultural division of labour was applied to the planet. As a result, peoples of distant zones were drawn into the vortex of change, the origins of which were obscure to them, while the European nations became dependent for their everyday activities upon a not yet ensured integration of the life of mankind. With free

trade the new and tremendous hazards of planetary interdependence sprang into being.[55]

In other words, while the production system is global, we have no means of integrating it with the life of mankind as a whole. Or rather, as Piketty points out, through numerous treaties, national states have since the 1980s deliberately relinquished the power to tax or regulate global companies in the interest of free circulation of capital, eroding the power of the welfare state to mitigate the deleterious effects of the production system.[56]

Although his focus is on land, Polanyi wants to include the whole of humanity's 'natural surroundings', and as we now know, that includes the oceans, the rivers, the atmosphere, the water we drink, and the air we breathe. Polanyi prefigures the work of environmental economists such as Herman Daly.

Daly marvels at the absurdity of the neoclassical model of the economy that leaves out the inputs from the environment and the output of waste to environmental sinks. What we need to think about, he argues, is how big the economy of production and consumption can grow relative to the capacity of the environment to sustain inputs of energy and materials and absorb waste. He questions:

> If we now recognise that growth is physically limited, or even economically limited in that it is beginning to cost more than it is worth at the margin, then how will we lift people out of poverty? The answer is painfully simple: by population control, by redistribution of wealth and income, and by technical improvements in resource productivity.[57]

The Market and Money

So far, we have seen how the protection of society and nature is necessary to the survival of capitalist production in a market society. Polanyi takes a further step in arguing that the production system itself requires protection from the self-regulating market. He explains:

> The need for protection arose on account of the manner in which the supply of money was organized under a market system. Modern central banking, in effect, was essentially a device developed for the purpose of offering protection without which the market would have destroyed its own children, the business enterprises of all kinds.[58]

This is the point of Joseph Schumpeter's 'theory of creative destruction' in which, in the process of competition for markets, older technology is constantly being replaced by newer technology – leaving older technology and its workforces stranded.[59] The advantage of creative destruction for increasing productivity and responding to new social needs, demands, and challenges is self-evident – even if new technologies bring their share of evils. But, in Polanyi's view, in that destructive process, businesses themselves must be protected through the stabilisation of money. Short-term instability of market prices under a system in which fixed contracts determined the price of labour and materials for businesses created a problem: 'Not low prices, but falling prices were the trouble'.[60]

Polanyi emphasises the distinction between commodity money and token money. The former is a produced commodity such as gold or silver that is chosen to function as money for the purpose of determining and comparing prices for various goods on international markets. Token money is money artificially created by national states through banknotes (or electronic signals) – worthless in themselves. Token money became necessary when the expansion of production and trade was not accompanied by an increase in the supply of commodity money (gold), resulting in price deflation.

Essentially, 'commodity money was vital to foreign trade; token money to the existence of domestic trade'.[61] Polanyi writes that, about the time of the Napoleonic wars (early nineteenth century), stable exchanges became essential to the very existence of the English economy. By centralising the supply of credit, it was possible to avoid the disruption to businesses and employment caused by unpredictable deflations. Credit restrictions and raising central bank interest rates were among the tools used. Interest rate increases, Polanyi observes, 'spread the effects of restrictions to the whole community while shifting the burden of the restrictions to the strongest shoulders'.[62]

Central banking meant that the automatic function of the gold standard in a self-regulating market 'was reduced to a mere pretence'. Central banks today still have the function of managing the deleterious effects of the market on either deflation or inflation. They still operate with essentially the same tools: controlling credit directly or through its price – interest rates. Central banking sustains the fiction of the separation of the political from the economic spheres.

The International Order

Polanyi begins and ends his narrative by writing about the international political order. The first chapters describe the 'hundred years' peace' amongst the European 'Great Powers' that reigned from the end of the Napoleonic wars until the beginning of the First World War. Trade was dependent on an international monetary system that could not function in a general war. International trade demanded peace, and the Great Powers were striving to maintain it.[63]

Changing political interests in society were involved. The middle classes, who had been a revolutionary force endangering peace up to Napoleonic times, now became the bearers of the 'peace interest'. The world economy was held together by international trade dependent on the gold standard. Polanyi writes, 'The breakdown of the international gold standard was the invisible link between the disintegration of the world economy which started at the turn of the century and the transformation of a whole civilization in the thirties'.[64]

Polanyi insists that the *political* function of the international monetary system has to be understood. That message resonates today as the international economy recovers incompletely from the global financial crisis, and neoliberalism has force-fed capitalism to the broken Soviet Union and its Eastern European minion states (see Chapter 5).

The outbreak of the Covid-19 pandemic has further destabilised the international economy. Central banks and national states have struggled to prevent first deflation and recession, and now inflation caused by the aftermath of the pandemic, the war in Ukraine, and businesses seizing the opportunity to raise prices – blaming external forces. Incomes of working people, already in decline for at least a decade, are now further reduced by rising interest rates. The poorest of society enter destitution while the top echelons of income and wealth balloon to ever greater proportions.

It would be foolish to make any assumption of a repetition of circumstances leading up to world war. The circumstances of the international economy are not the same as in the 1930s. The 'peace interest' has been maintained for 77 years not only by global trade but by the horrifying consequences of nuclear war. Yet, there are certain political similarities.

Writing of 'history in the gear of social change', Polanyi discusses the rise of both fascism and socialism, and the facing off between the two forces and their political actors in the 1930s. Very little in the way of political ideology or local culture, he argues, actually connects the outbreaks of fascism at that period.

So varied were these circumstances that the appearance of fascism in the industrial countries cannot be ascribed to local causes, national mentalities, or historical backgrounds: 'Fascism had as little to do with the Great War as with the Versailles Treaty, with Junker militarism as with the Italian temperament. The movement appeared in defeated countries like Bulgaria and in victorious ones like Jugoslavia, in countries of Northern temperament like Finland and Norway and of Southern temperament like Italy and Spain' – in countries of Arian and non-Arian race, and both Catholic and Protestant religion.[65]

It is as well to remember Polanyi's inclusive definition of fascism when we later discuss in Chapter 8 the more restrictive definition of Jan-Werner Müller.[66]

The symptoms that led to the rise of fascism were various:

the spread of irrationalistic philosophies, racialist ethics, anticapitalistic demagogy, heterodox currency views, criticism of the party system, widespread disparagement of the 'regime', or whatever was the name given to the existing democratic setup.[67]

All these symptoms are readily identifiable today, for instance in the USA, Britain, and France.

Moreover, fascist movements of the 1930s derived their strength not so much from popular manifestations and mass followings as from the influence of the people already in positions of power. Again, that feature is today manifest in the USA in the persons of Donald Trump and most of the Republican Party, in Hungary in the person of Viktor Orban, in Turkey in Recep Tayyip Erdogan, and in Russia in Vladimir Putin and his United Russia Party.

As to socialism, Polanyi observes that the Russian revolution really consisted of two periods: the first from 1917 to 1921 followed the pattern of European revolutions from Cromwell's Commonwealth of 1653 in England to the French Revolution of 1789 – revolts of sundry dispossessed classes against autocratic regimes. The second period, with the collectivisation of agriculture, saw the start of a socialist transformation that 'transformed our world in the thirties'.[68] 'For the first Russian Revolution achieved the destruction of absolutism, feudal land tenure, and racial oppression – a true heir to the ideals of 1789 [the French Revolution], the second Revolution established a socialist economy'.[69]

From Polanyi's perspective of the 1940s, Russian communism was a success in transforming an agrarian empire into a modern industrial economy without the trauma of rapid imposition of market society and with the socialisation of land. Polanyi might not have known about the famines caused by collectivisation, and certainly did not know of the scale of Stalin's purges. There is certainly no doubting the influence of communism in Russia on Western social philosophy. But, as Hobsbawm points out, 'The possibility of dictatorship is implicit in any regime based on a single irremovable party. In a party organized on the centralized hierarchical basis of Lenin's Bolsheviks, it becomes a probability'.[70]

Freedom in a Complex Society

The questions at the heart of Polanyi's narrative of transformation are these: How can social protection be reconciled with capitalist production in a market society? How can individual freedom be guaranteed in a society that ensures protection? Both questions are as relevant today as in the times of Polanyi's historical analysis. Moreover, as has become very clear today (as Polanyi understood), we have to think not about Humanity *and* Nature, but Humanity *in* Nature – indivisibly so.

In relation to the first question, Polanyi points to the disintegration of nineteenth-century society, which, structured as it was by liberal ideology, failed to provide an answer to that question.

> Apart from exceptional circumstances such as existed in North America in the age of the open frontier, the conflict between the market and the elementary requirements of organized social life provided the century with its dynamics and produced the typical strains and stresses which ultimately destroyed that society.[71]

The failure of liberalism arose from its vision of human social behaviour. Humans in their natural state were portrayed as isolated individuals pursuing their material self-interest. In their economic behaviour, humans would tend to abide by what was described as economic rationality. If only they could be governed in a way that allowed that *natural* condition to prevail, all would be for the best in the best of all possible worlds. The only *social relations* that were required were the social relations of the market. The only *government* that was required was government of and by the market. The only national *sovereignty* was that guaranteed by the possession of central banks. The only *protection* required was the protection of the right of ownership of property.

Such an organisation of economic life, Polanyi argues (drawing on anthropological studies), is entirely unnatural, in the strictly empirical sense of being exceptional. The market organisation of society was the outcome of a conscious and often violent intervention on the part of government, which imposed the market on society for noneconomic ends. It was also, in the long run, completely unsustainable.

From the 1930s on, societies across the world were transforming rapidly. Both through practical steps letting go of the liberal dogma and through rethinking the nature of freedom and the limits of control. Across Europe, the needs of society were prioritised over the market in a variety of ways by different kinds of regime: democratic, aristocratic, constitutionalist, and authoritarian. The commodity fictions of labour, land, and money were 'disestablished' in all directions of the social compass.

Hobsbawm writes of the profound effect of the world economic breakdown between the two world wars. Indeed, 'the world of the second half of the twentieth century is incomprehensible without understanding the impact of the economic collapse'.[72] Thus, began a process of world-wide social transformation of extraordinary rapidity and universality: decolonisation, the virtual disappearance of the peasantry, mass urbanisation, and the development of industries requiring an educated workforce.

In this transformation (as radical as the establishment of a competitive labour market), the wage contract ceased to be a private contract except on subordinate and accessory points. 'Not only conditions in the factory, hours of work, and modalities of contract, but the basic wage itself, are determined outside the market'.[73] What we now know as town and country planning – though Polanyi did not use the term – removed land from the market, designating areas for different land uses. Control of money was subject to government direction of investment and regulation of the rate of saving.[74]

This social transformation in the democracies involved two important features that Polanyi observed. First, as he writes:

> The nature of property, of course, undergoes a deep change in consequence of such measures since there is no longer any need to allow incomes from the title to property to grow without bounds, merely in order to ensure employment, production, and the use of resources in society.[75]

Second, the end of a market society does not mean the absence of markets. 'These continue, in various fashions, to ensure the freedom of the consumer, to indicate the shifting of demand, to influence producers' incomes, and to serve as an instrument of accountancy, while ceasing altogether to be an organ of economic self-regulation'.[76]

As Polanyi reaches the conclusion of his narrative, he approaches the philosophical questions of power and freedom in societies. Discarding the market utopia, he states, 'brings us face to face with the reality of society'.[77] It is an illusion, Polanyi argues, to conceive of a society shaped by the desire and will of individuals alone, 'which equated economics with contractual relationships, and contractual relations with freedom'.[78]

No society is possible, he posits, in which power and compulsion are absent. If we follow Piketty, we can agree that no society is possible without *ideology* to explain and justify inequality. The liberal illusion is simply the ideology with which the imposition of market relations was allowed to transform society. There was, in that transformation, no absence of state power, as Polanyi has amply demonstrated in his narrative.

Facing the reality of society clarifies the dividing line between liberalism, on the one hand, and fascism and socialism on the other. Liberalism is constituted not by the reality of actual societies but by a mythical and theoretical utopia.[79] Fascism and socialism both stem from the recognition of actual societies: 'the reality of society is accepted with the finality with which the knowledge of death has molded (sic) human consciousness'.[80]

What separates socialism and fascism is ethical (or as Polanyi puts it 'moral and religious'), not economic. The difference hinges on the nature of freedom conditioned by power. In the final paragraphs of his book, Polanyi expresses the stark choice facing actual societies: 'The discovery of society is thus either the end or the rebirth of freedom. While the fascist resigns himself to relinquishing freedom and glorifies power, which is the reality of society, the socialist resigns himself to that reality and upholds the claim to freedom in spite of it'.[81]

In the twentieth century, socialism was ushered into capitalist societies enacted in many different ways in different nations: in the Nordic countries and the Netherlands, in Britain and France, in pre-war and post-war Germany, and at different times in Spain and Portugal.

Thomas Piketty describes the social democratic transformation in the first half of the twentieth century as 'The Great Redistribution' across much of Northern Europe.[82] Britain, the country which gave birth to liberalism (and yet until 1909 governed by the House of Lords), gave the Labour Party in 1945 an overwhelming majority in the House of Commons, which created the social welfare insurance system, the national health service, greatly expanded compulsory public education, the nationalisation of power and public transport, and the planning of land use and development.

Polanyi's narrative leaves us before the socialist transformation in Britain. In the following chapters, we turn from the long history of *The Great Transformation* to more immediate and politically contingent transformations. First, the transformation wrought by the British Labour Government from 1945.

Notes

1 Including critical theorists such as Gramsci, Foucault, Habermas as well as many lesser figures, I do not mean to discount the contribution of founders of sociology such as Max Weber and Emil Durkheim.
2 Polanyi, K. (1944) p. 3.
3 Camilleri, J. (1995) pp. 207–228.
4 Ibid., p. 43.
5 Piketty, T (2020) p. 175.
6 Ibid., p. 175. Country folk had taken to blackening their faces during night-time poaching in order not to be seen or caught.
7 Polanyi, K. (1944) pp. 81—82.
8 Ibid., p. 83.
9 Ibid., p. 85.
10 The poor consisted simply of the common people: those who were without a private income to keep them in leisure.
11 By for instance Friedrich Engels: *The Working Class of England in 1844.*
12 Polanyi, K. (1944) pp. 163–164.
13 Ibid., p. 79.
14 Introduction to *The Great Transformation* by Fred Block. Polanyi, K. (1944) p. xxii.
15 Polanyi, K. (1944) p. 76.
16 Ibid., p. 88.
17 Polanyi K. (1944) p. 118. Citing Townsend's *Dissertation on the Poor Laws* (1787).
18 Ibid., p. 119.
19 Piketty, T. (2020) Chapter 6.
20 Polanyi K. (1944) p. 122.
21 Ibid., p. 122: footnote: Bentham, J. *Principles of the Civil Code*. Chapter 4, Browning, Volume 1, p. 333.
22 Ibid., p. 122.
23 Ibid., p. 129. Smith, A. *Inquiry into the Nature and Causes of the Wealth of Nations.*
24 Ibid., p. 129.
25 Piketty, T. (2022) p. 16 and Figures 1 and 2.
26 Polanyi, K. (1944), p. 110.
27 Ibid., p. 111.
28 Drawn from 'Robert Owen and New Lanark' (https://spartacus-educational.com/ ExamIR3.htm. Accessed 21/07/2022).
29 Polanyi (1944), p. 134.
30 Ibid., p. 135.
31 Ibid., p. 175.
32 Ibid., p. 180.
33 The six points of the People's Charter called for: a vote for every man aged 21 years and above, of sound mind, and not undergoing punishment for a crime; the secret ballot to protect the elector from coercion or bribery; the abolition of the property qualification for membership of Parliament; payment for members of Parliament to provide for persons without private incomes to attend; constituencies of equal population 'securing the same amount of representation for the same number of electors'; and annual parliamentary elections as a further check on bribery and intimidation. https://en.wikipedia .org/wiki/Chartism (accessed 29/07/2022).
34 Polanyi, K. (1944) p. 181.

35 Ibid., p. 180.

36 Ibid., p. 138.

37 Ibid., p. 162.

38 Ibid., p. 163.

39 Ibid., p. 159.

40 Ibid., p. 162.

41 Ibid., p. 151.

42 Ibid., Chapters 12 and 13.

43 Ibid., p. 153

44 Ibid., p. 156.

45 Ibid., p. 148.

46 I have not altered Polanyi's language to accommodate feminists' proper insights on language. By 'Man' of course he means People.

47 Or perhaps both at once! Polanyi, K. (1944) p. 164.

48 Ibid.

49 Ibid.

50 Ibid., p. 165.

51 Ibid., p. 186.

52 Ibid., p. 75.

53 Ibid., p. 193.

54 Ibid. p. 189. '*Thus, the mobility of goods to some extent compensates the lack of inter-regional mobility of the factors*' (citing B. Ohlin, 1935, *Interregional and International Trade*. Author's italics).

55 Ibid., p. 190.

56 Piketty, T. (2022) p. 170.

57 Daly, H. (1996) p. 7.

58 Polanyi, K. (1944) p. 201.

59 Schumpeter, J.A. (1943) Schumpeter showed that substitute mechanisms analogous to markets and prices were feasible in a planned economy.

60 Polanyi, K. (1944) p. 201.

61 Ibid., pp. 202–203.

62 Ibid. p. 203. Today, with central banks' post-pandemic rate hikes designed to curb inflation, the impact does not appear to burden 'the strongest shoulders', rather falling on those seeking home ownership (see Chapter 10).

63 Ibid., p. 16.

64 Ibid., p. 21.

65 Ibid. pp. 245–246.

66 Müller, J-W (2022).

67 Polanyi (1944), p. 246.

68 Ibid., p. 255.

69 Ibid., pp. 255–256.

70 Hobsbawm, E. (1995) p. 389.

71 Polanyi (1944), p. 257.

72 Hobsbawm, E. (1995) p. 86

73 Polanyi (1944), p. 259.

74 Ibid., pp. 259–260.

75 Ibid., p. 260.

76 Ibid., p. 260.

77 Ibid., p. 267.

78 Ibid., p. 266.

79 Anyone reading the masters of classic and modern liberalism cannot but be struck by the abstract logics and the absence of empirical evidence of actual societies. Mainstream liberal economics remains to this day an exercise in logic based on simplistic and untested assumptions about human behaviour.

80 Polanyi, K. (1944) p. 267.
81 Ibid., p. 268.
82 Piketty, T. (2022) p. 127.

References

Camilleri, J. (1995) 'State, civil society, and economy', in Camilleri, J., Jarvis, A.P., and Paolini, A.J. eds. *The State in Transition, Reimagining Political Space*, Boulder, CO: Lynne Rienner Publishers, p. 214.

Daly, H. (1996) *Beyond Growth, The Economics of Sustainable Development*, Boston: Beacon Press.

Hobsbawm, E. (1995) *The Age of Extremes 1914–1991*, London: Abacus.

Müller, J.-W. (2022) *Democracy Rules*, London: Penguin Books.

Piketty, T. (2020) *Capital and Ideology*, tr. Arthur Goldhammer, Cambridge: Bellknap Press of Harvard University.

Piketty, T. (2022) *A Brief History of Equality*, tr. Steven Rendall, Cambridge: Bellknap Press of Harvard University.

Polanyi, K. (1944, 2001 edition) *The Great Transformation, The Political and Economic Origins of Our Time*, Boston: Beacon Press.

Schumpeter, J.A. (1943) *Capitalism, Socialism and Democracy*, London: Allen and Unwin.

3

THE ARRIVAL OF SOCIAL DEMOCRACY IN BRITAIN

Introduction

Today, in a world dominated by liberal 'free market' ideology, it is hard to imagine the world of the welfare state in Britain after 1945, a world which survived with modifications until the end of the 1970s. It is hard to imagine the cradle-to-the-grave social protection: social services for all, free education to the secondary level and large-scale assistance for tertiary level, free health care for all, insurance against unemployment, generous old-age pensions, and strong rights for the working people. It is hard to imagine the key elements of the economy being nationalised, and rights to develop land taken over by the state, in order to create town and country planning.

It is hard to imagine the degree of progressive taxation of income and wealth, the abandonment of the liberal ideology of sanctification of property, the fall of inequality and the rise of social opportunity. The institutions that were created had flaws, and the social democratic project was far from perfect. But Polanyi, writing in the 1940s, believed that free market liberalism was finished for ever.

The ideology of liberalism was to return in the 1980s, but something of social democracy stubbornly remained because new public expectations of government had been generated, setting in motion again the intractable conflict between protection for humans and nature and the principle of the market.

We will look first at the structural crisis that preceded this 'social democratic' moment in Britain, then at the change of ideology and the champions who brought about the culmination of social democracy. Finally, we will consider some of the consequences.

DOI: 10.4324/9781003382133-4

Crisis

The period from 1914 to the 1930s was a time of war, followed by mass unemployment, industrial turbulence, and popular mobilisation against capitalism. During the First World War, some 6 per cent of the adult male population of Britain were killed in the fighting. To make matters worse, the flu pandemic from 1917 to 1919 left a legacy of horror, particularly for the overcrowded poor in the industrial cities.

Following a brief depression at the end of the war in 1921, some 23 per cent of the British workforce was left unemployed. Inspired by the success of Leninism in Russia, revolutionary movements broke out across Europe demanding an end to capitalism. The British Communist Party, formed in 1922, set about organising the unemployed and taking leading roles in some of the trade unions. In November 1922, unemployed workers marched from the North of England, Scotland, and Wales to London. Some 70,000 workers descended on Parliament.

Unemployment declined after 1923 but still hovered around 10 per cent. In 1926, the General Council of the Trades Union Congress called a general strike across all heavy industries including, especially, transport and mining. Some 1.7 million workers went on strike to prevent the government from imposing wage reductions and worsening working conditions. From 1929 to 1932, the Great Depression hit Britain, bringing a renewed increase in unemployment to more than 20 per cent. The meagre and means-tested unemployment benefits left many in dire poverty.

The Second World War, unlike the First World War, created an existential crisis for Britain. After the withdrawal from Dunkirk it was clear that this war would not be a continental war in which Britain would participate, but a war waged *on* Britain and probably *in* Britain itself: a war of territorial defence as much as overseas military expeditions. To win such a war required a colossal collective effort.

As the historian Tony Judt remarks, for the countries under Nazi occupation the war was primarily a civilian experience, of repression, exploitation, and extermination. In Britain it was a civilian experience of a different kind, an experience of bombardment, blockade of overseas supplies, and shortage of food. In order to fight and win the war the British ransacked their own resources: by the end of the war, Great Britain was spending more than half its Gross National Product on the war effort.[1]

It was also a war in which the fear of communism had to be put aside. After Nazi Germany attacked Russia on a long front from the Baltic to the Black Sea, the Red Army fell back from one defended line to another as far as Stalingrad, at which point Stalin ordered an end to retreat. The occupation and fightback took a terrible toll on the civilian population and soldiers alike. Some 16 million Soviet civilians were killed in death camps and killing fields from Odessa to the Baltic. Some 5.5 million Soviet soldiers were captured, of whom 3.3 million died from starvation, exposure, and mistreatment in German camps. All told, the USSR lost around 10 per cent of its pre-war population.

The Soviet engagement on Germany's eastern front undoubtedly saved Britain from invasion and ultimately led to Germany's defeat. Recognition of the Soviet role should not in any way diminish the human cost of the British and American war effort. However, the cost to Russia in human sacrifice went largely unrecognised in the heroic post-war narratives of Britain and America, something that left a rancorous impression on Russian governments and peoples ever since. The Russian fight back against the Axis powers was conducted with a ferocity that matched that of the Wehrmacht and perhaps left a lasting impression on how wars were to be fought – evidenced in subsequent Russian wars in Chechnya, Georgia, Syria, and Ukraine.[2]

Mobilising for the Second World War, 'Britain was transformed from a primarily free market economy to a planned economy'.[3] The output of the war economy was directed and managed by central ministries with wide powers of intervention such as the Ministry of Aircraft Production and the Ministry for War Transport. The entire population was engaged. Milward records that Britain had a highly successful record of mobilising the home front: maximising output, assigning the right skills to the right task, and maintaining the morale and spirit of the people.[4] A similar popular mobilisation of support occurred in Roosevelt's America.[5]

The war effort was therefore not just a matter of mobilising industry but of gaining the support of the whole population: The war was a 'people's war' that enlarged democratic aspirations and produced promises of a better Britain after the war. The people's war also 'signified the popular demand for planning and an expanded welfare state'.[6]

Judt, in his post-war history, corroborates this conclusion for Western Europe generally. He argues that the welfare states were not politically divisive:

> They were socially redistributive in general intent (some more than others) but not at all revolutionary – they did not 'soak the rich'. On the contrary: although the greatest immediate advantage was felt by the poor, the real long-term beneficiaries were the professional and commercial middle class.[7]

Their members had access to free health, unemployment, and retirement benefits; along with state-provided free or subsidised secondary and higher education for their children, they left the salaried and white-collar classes with a better quality of life and increased disposable income.[8]

Most importantly, the war created a sense of solidarity among the people of Britain. With the air raids on British cities, the civilian population was now in the front line of battle. Culturally the significance of class differences diminished. Even in the armed forces the distinctions of rank faded a little. With conscription, Britain now depended on a citizen army. The army authorities understood that working-class soldiers needed to be treated with respect and their material and recreational needs met. The Navy, Army, and Airforce Institute (the NAAFI) was founded to provide leave camps equipped with swimming pools, gardens, shops, restaurants, dance halls, and bars.

There was a widespread sense that society could not be allowed to revert to its pre-war state with the desperation, poverty, and unemployment of the Depression years. As Addison observes, a new test of patriotism 'was therefore to be loyalty to the needs and aspirations of the people of Britain: social patriotism, as it may be called': the sense that 'men and women fighting for their country were entitled to a fair share of the fruits of victory in accord with natural justice'.[9]

Judt recalls 'the social democratic moment':

> In the peak years of the modern European welfare state, when the administrative apparatus still exercised broad-ranging authority and its credibility remained unassailed, a remarkable consensus was achieved. The state, it was widely believed, would always do a better job than the unrestricted market: not just in dispensing justice and securing the realm, or distributing goods and services, but in designing and applying strategies for social cohesion, moral sustenance and cultural identity.[10]

This new sense of justice, a cultural transformation extending across class, race, and gender paved the way for collective welfare to become the foundation of ideological and institutional change.

Ideology

As Polanyi recorded, from the late nineteenth to the early twentieth century the clash between liberal ideology and the need to protect humans and nature from the negative effects of the market turned in favour of protection. Piketty notes that the last gasp of the sanctification of property in France occurred when that country imposed crushing reparations on Germany in the aftermath of the First World War – 'at a rate of 300 per cent of the German national income at a time when the country was already on its knees'.[11] This crippling tax is well understood to have nourished the ideology of Nazism.

In the twentieth century, however, there was a further development: beyond 'protection' to what Piketty terms 'the march towards equality'.[12] The battle for progressive taxation in Britain followed hard on the heels of the battle for universal suffrage and the secret ballot. Electoral reforms in 1867 and 1884 established standard voting rules across the nation. Universal male suffrage was established in 1918, and the vote finally extended to women in 1928.

As electoral democracy grew, so did the influence on governments of classes otherwise excluded from power, first the middle and then the working class. The former Whig Party was renamed the Liberal Party and began to take up the cause of the new voters and 'adopt a platform and ideology much more favourable to the middle and working class'.[13]

In effect, the new Liberal Party split from the proprietarian ideology of classical liberalism. Underlying the political change was a change in the way property was conceived. The belief faded that monetary property – income and wealth – was an absolute right without which a market economy was impossible. The British House of Lords tried to hold back the change – hardening, protecting, and making sacred the right of property.[14] But before the First World War, the Liberal Government, fearing the advance of the Labour Party, decided to propose measures to help the working class, including pensions for workers.

The 'People's Budget' of 1909 (introduced by Chancellor of the Exchequer David Lloyd George allied with the young Winston Churchill – then a Liberal) combined strongly progressive taxation, including a land tax (following the ideas of the American reformer Henry George) with a limited range of social welfare programmes.[15] When the budget was opposed by the Lords, the government introduced a new bill in the House of Commons to stop the Lords from vetoing or amending finance bills and limiting their power to block other legislation, in effect changing the bundle of precedents and conventions that compose the British Constitution.

Under threat of stacking the Lords with hundreds of new members, the Lords capitulated in 1911. From 1911, then, the House of Lords, as Piketty observes, 'forfeited all legislative power ... it is the will of the majority as expressed in the ballot box and in the House of Commons that has the force of law in the United Kingdom'.[16] The House of Lords, the final bastion of the absolute right to property, was reduced to a purely consultative and largely ceremonial role.

Meanwhile, the Labour Party grew in representation in the House of Commons, rising to minority government in 1924 and from 1929 to 1931 under Ramsay MacDonald. The Party espoused national economic planning under Clause IV of the party constitution drafted by Sydney Webb (co-founder of the London School of Economics):

> To secure for the workers by hand or by brain the full fruits of their industry and the most equitable distribution thereof that may be possible upon the basis of the common ownership of the means of production, distribution, and exchange, and the best obtainable system of popular administration and control of each industry or service.[17]

The socialist ideology of the Labour Party, it should be noted, was at that time strongly influenced by Marx and to some degree by the Leninist revolution in Russia – though always retaining parliamentary democracy. Equitable distribution was to be achieved partly through state control of the means of production.

In Sweden, the influence of Marx on economic policy was also strongly felt in the social democratic movement, but for the 1932 programme that brought the Social Democratic Labour Party (SDLP) to power, one of the leaders of the Party, Ernst Wigforss, argued that the SDLP had two roots: Marxism and economic liberalism. 'The socialist world view', he said, 'contains two lines of thought that do not necessarily coincide: the demand for nationalization and the demand for a planned economy. Nationalization is not necessarily the same as economic planning'.[18] Economic planning did not require nationalisation. 'Therefore, it should be possible to lead production and keep it going by a number of means than giving rights of ownership to society'.[19]

Over the following years from 1932 to 1976 when the Party (now the Social Democratic Party) was continuously in power, the Party developed an ideology that embodied a concept of democracy in which the democratic ideal should infuse not only political life but social and economic organisation as well. The sociologist Gustav Steffen wrote that 'deproletarianization' should be the aim: the liberation of workers from being a 'raw material used by others', lacking property and the opportunity to participate fully in society'.[20] These ideas led the SDP in a direction different from the British Labour Party, enabling the Party to expand equality via an immense enlargement of universally available social services without some of the confrontational politics that beset the British Labour Party in later years.

Swedish social democracy forged an ideology that included the whole population. Per Albin Hansson proposed the leading idea of *folkhemmet*, society and state working together to create 'the people's home'. This would mean 'breaking down the barriers that divide citizens into privileged and misfavoured, into rulers and dependents, into rich and poor, the glutted and the destitute, the plunderers and the plundered'.[21] Tilton writes that, 'solidarity and equality of consideration characterize the good home; there consensus is the objective and democratic persuasion the method of governance'.[22]

As we have shown above, external circumstances and crises, social movements, public demonstrations and protests, ideologies and institutions are involved in the process of transformation. Piketty sums up the situation:

> everything depends on the rules and institutions that each human society establishes, and things can change very quickly depending on the balance of political and ideological power among contending social groups as well as the logic of events and on unstable historical trajectories, which can be understood only through detailed study.[23]

Ideology is about more than ideas and institutions. It includes feelings of justice and injustice deeply embedded in a nation's culture, the *political ethos* as I will call it. As noted above, the war changed those feelings throughout the population of Britain. But we should not overlook the role of actors in tipping their world into a profound institutional transformation. Democracy empowers such actors, but they require persistence, commitment to their vision, and preparedness to take political risks. We turn to some of those actors whom we term 'champions', those at the peak of state power who shaped the welfare state in Britain.

Champions

From May 1940 to May 1945, Winston Churchill and Clement Attlee led their respective Conservative and Labour parties into a national coalition to conduct the Second World War against the Axis powers. Initially Churchill formed a war cabinet with Neville Chamberlain (Churchill's predecessor as Conservative Prime Minister) as Lord President of the Council, Attlee as Lord Privy Seal, Viscount Halifax (Conservative) as Foreign Secretary, and Arthur Greenwood (Labour) as Minister Without Portfolio.[24]

The keeper of the monarch's private seal is an obsolete function, but like the Lord Chancellor (who is responsible for justice and the courts of law) and Lord President of the Council (responsible for the monarch's Privy Council), the Lord Privy Seal is one of the great offices of state. In this instance the titles surely meant that the holders had more or less equal power in a Cabinet (with the PM the first among equals) before the office holders' actual functions could be sorted out.[25]

The Cabinet membership was quickly increased to eight with Ernest Bevin (Labour: Minister of Labour and National Service), Anthony Eden (Conservative, replacing Halifax: Foreign Secretary), Lord Beaverbrook (formerly Max Aitken, publisher of the Daily Express: Minister of Aircraft Production), Sir Kingsley Wood (Conservative: Chancellor of the Exchequer – Treasurer), and Sir John Anderson (an eminent scientist and public servant, Lord President of the Council replacing Chamberlain who died in November 1940).

The Cabinet was gradually enlarged to include Labour Party leaders: Stafford Cripps and Herbert Morrison. Attlee became Deputy Prime Minister. Churchill essentially handed Attlee responsibility for the home front while he himself, with his military general staff, attended to the conduct and rhetoric of the war. Churchill trusted Attlee, who had strongly opposed appeasement of Hitler, despite (and perhaps because) allegedly calling Attlee 'a modest little man with much to be modest about'. The formation of the national coalition was also a recognition of the cultural sea change of social patriotism, embodying a longing for social justice.

Attlee and his Labour colleagues (within and outside Cabinet) were preparing for the implementation after the war of the Beveridge Report that had been commissioned by Arthur Greenwood (Minister without Portfolio) in 1940: 'To undertake, with special reference to the inter-relation of the schemes, a survey of the existing national schemes of social insurance and allied services, including workmen's compensation, and to make recommendations'.[26]

William Beveridge was a social economist and (formerly) a progressive Liberal politician. He worked with the mathematician Jane Beveridge, his wife. Beveridge had studied the problem of unemployment and experienced its results in his work as sub-warden of Toynbee Hall, a 'settlement house' sheltering people in poverty and homelessness.[27] The socialist Beatrice Webb had introduced Beveridge to Churchill, who invited him to serve as an advisor to the Board of Trade.

Beveridge had earlier argued that the way industries were organised was responsible for unemployment.[28] Influenced by Keynesian economics, he believed that social modification of capitalism was preferable to state socialism on Marxist lines.[29]

He upheld three principles in his report. First, that the future should not be determined by sectional interests and any solution to contemporary problems must address the big picture – there must be no 'patching up' of problems. Second, that social insurance is just one element in a comprehensive policy of social progress. Third, that social security must be achieved by cooperation between the state and the individual. The state should provide a national minimum that would ensure the opportunity for further improvement through individual effort.

Beveridge identified five main evils of capitalist society, colourfully describing them as 'giants' to be slain: want, disease, ignorance, squalor, and idleness. To defeat the 'giants' a comprehensive system of universal (non-means tested) welfare was recommended. This included insurance against unemployment financed by employees and employers through payroll deductions, a free health service, family allowances, disability benefits, age pensions, and workers' compensation.

In 1943 and 1944 the government issued a series of reports (White Papers) 'each one a victory for the massed ranks of social democracy'.[30] The paper, '*Educational Reconstruction*' (1943) proposed free secondary education for all and laid the foundation of the Education Act (1944). Free and comprehensive health care for all, across every branch of medicine was put forward in *A National Health Service* (1944). *Employment Policy* (1944) set out the duty of the state to maintain a high and stable level of employment. *Social Insurance* (1944) dealt with the government's adoption of Beveridge's national insurance scheme to cover pensions and unemployment benefits. Only *Housing Policy* (1945) was, in Addison's assessment, 'a vague and feeble document of barely eight pages with no clear target for the number of homes to be built in the post-war period'.[31]

Beveridge's report was not the only one to influence the Labour leadership as preparations were made to offer the country a socialist alternative. The report of the Royal Commission on the Distribution of the Industrial Population chaired by Sir Montague Barlow observed how employment in traditional heavy industries located close to natural resources was being replaced by the new 'local' industries close to markets in Birmingham and the London region.[32]

The economic advantages of such locations were accompanied by disadvantages such as high site values, traffic congestion in the very large towns, and long daily journeys between home and workplace incurring fatigue, discomfort, and expense. Thus, urban and regional planning was required both to deconcentrate the new industries, assist depressed areas of declining industry, and improve the environment of the new workforce in the Midlands and the South East. The report addressed questions of town planning, garden cities, satellite towns, and trading estates, and restrictions on the further outward sprawl of London, supporting the need for new planning institutions.[33]

Town planning in Britain has a long history, as described by Cullingworth who writes: 'The Barlow Report (1940) is of significance not merely because it is an important historical landmark, but also because some of its major recommendations were for long accepted as a basis for planning policy'.[34]

The Scott Report addressed the problems of rural England and protection of the countryside: a place of 'splendour and leisure', and one that should be safeguarded for future generations. The Uthwatt Report set out measures relating to compensation and betterment resulting from state intervention. These three reports became the foundation of the town and country planning systems in England and Scotland.

The champions of welfare state socialism were a mix of the well-educated, professional middle-class, and working-class political leaders with links to the trade unions. They were backed by, and inspired by, a powerful mass working-class movement, but they also appealed to the values of middle-class voters.

Attlee, the son of a wealthy London solicitor, was educated at Haileybury College (an English public school) and Oxford University. Arthur Greenwood, the son of a painter and decorator, was educated at the Yorkshire College (subsequently becoming the University of Leeds). Ernest Bevin, one of the founders of the Transport and General Workers' Union had a modest education in Devonshire. Hugh Dalton was born in Wales, the son of a clergyman and chaplain to Queen Victoria. He was educated at Eton College and King's College Cambridge.

Aneurin Bevan, who took on the implementation of the National Health Service, grew up in a Welsh working-class family – the son of a coal miner. Emmanuel Shinwell was born in London's East End in a large family of Jewish immigrants. He moved to Glasgow and became a union organiser on 'Red Clydeside'. Herbert Morrison grew up in Stockwell, the son of a police constable, and left school at the age of 14 years, elected as a Labour Councillor for the Borough of Hackney in 1919.

A transformation requires actors: individuals with access to the power of the state are able to change the institutions governing the lives of the people and the relationship of people to the natural environment. But no 'champion' acts alone. There are leaders and supporters, and they in turn must have the support of more actors and movements in the community. That the Labour Party was able to win power in 1945 was surely greatly assisted by its leaders already having been brought into government by Churchill during the war, and by the political pressure exercised by the power of the trade union movement of the time.

Consequences

The transformation of British society and economy wrought by the Attlee Government was rapid, comprehensive, and dramatic. In the 1945 general election won by the Labour Party, the new government did not hesitate to set forth a democratic–socialist programme – one that would endure with modifications for nearly 40 years.

The Party's social justice agenda was delivered in just 6 years from 1945 to 1951. The immediate aftermath of the war saw a period of severe austerity in which the government set about replacing the financial resources destroyed or used during the war. Food rationing, initiated during the war, was continued. Taxes were raised on the highest incomes and inheritances to very high levels.

The socialist programme generated full employment, avoided a return to Depression, gave rise to a period of nearly 30 years of economic growth, and successfully resettled the soldiers, sailors, and air-force personnel into society.

It is important from the perspective of today – in a very different ideological world – to be clear about exactly what was achieved. The programme included: nationalisation of the key elements of industry, the creation of the welfare state, the institutionalisation of urban and environmental planning, industrial law reform, and the beginning of decolonisation.[35] In summary, the following were the main elements of the programme.

Nationalisation

The core sectors of production, communication, and finance were brought under state ownership. The entire coal industry was nationalised under the control of the National Coal Board. The Transport Act (1947) created the British Transport Commission which brought the four main railway companies under state ownership to form 'British Railways'. Inland water transport and some road passenger and haulage firms were brought under the British Transport Commission (becoming 'British Road Services', 'British Waterways', and including even the travel agency Thomas Cook and Son).

Electricity and gas production and supply were nationalised under the Electricity Act (1947) and the Gas Act (1948). Local electricity and gas boards were created to manage supply. The Bank of England was nationalised in 1946 and remains a government-owned agency, albeit with a degree of independence (though the government appoints commissioners). The iron and steel industries were nationalised becoming the Iron and Steel Corporation (Iron and Steel Act, 1949).

The Welfare State

The National Insurance Act (1946) provided sickness and unemployment benefits and retirement pensions. The National Assistance Act (1948) set up a safety net for anyone not covered under national insurance. Hospitals were brought under state control and universal free health care was introduced with the National Health Act (1946). The National Health Service created publicly funded healthcare covering all healthcare needs, including optical and dental treatment. Consultants were paid salaries that provided a good standard of living without their having to become private practitioners.

Planning, Housing, and Countryside

Funding for a programme of expanded public housing was provided, to be constructed by local authorities. The New Towns Act (1946) led to comprehensive plans for the growth of new towns to relieve overcrowding in the major cities, especially London and Glasgow. The Town and Country Planning Act (1947) nationalised the right to develop private land, providing a fund of £300 million to compensate landowners for loss of their development rights. The Act also granted the right to local authorities to purchase land compulsorily where the public interest required it. Grants were provided by the government to local authorities for redevelopment.

The National Parks and Access to the Countryside Act (1949) designated landscapes as 'National Parks' and 'Areas of Outstanding Natural Beauty' and provided for the recording, protection, and maintenance of public footpaths through private land.

Powerful local planning authorities were created to control land use and protect agricultural land. Regional offices were set up within the Ministry of Town and Country Planning to lead regional-scale development. Comprehensive Development Areas were designated giving local government compulsory purchase powers to redevelop areas suffering from urban decay and war damage.

Law Reform, Industrial Relations, and Workers' Rights

Laws were introduced to improve workplace conditions, provide for sick pay and workers' compensation for injury suffered at the workplace. New pension schemes and retirement benefits were created for various industries: fire services, electricity, and merchant shipping. Breaks from work during the working day were legislated: lunch breaks of at least 45 minutes and tea breaks of 30 minutes for people working between 4.00 pm and 7.00 pm. A Fair Wages Resolution required employers given government contracts to give workers the right to join trade unions. The Agricultural Wages Board ensured fair wage levels and accommodation for workers on the land. Health and safety regulations were imposed across industries involving work with hazardous materials such as mining, metal processing, blasting, casting, and pottery which generated dust or fumes.

The Criminal Justice Act (1948) imposed new humanitarian conditions on the punishment of offenders, abolishing corporal punishment, hard labour, and penal servitude. Laws were introduced to enable employees to sue their employers in cases of injury due to negligence of a fellow worker, and provisions were made for legal aid for those who could not afford legal fees.

The International Situation

The government played an active part in the United Nations, and strongly supported the new British Commonwealth. Even with financial assistance in grants and loans from the USA and Canada, it was clear that Britain's imperial rule could not continue. After the war, Britain was on the verge of bankruptcy. Maynard Keynes had warned that the country could not afford the massive expenditure on the colonies and global military deployment. The government withdrew its forces from around the world as quickly as possible. The Indian sub-continent was rapidly partitioned (causing untold hardship and strife in the process). Independence was granted to India and Pakistan, Ceylon (Sri Lanka), and Burma (Myanmar).

Piketty explains that in addition to the rise of the welfare state and progressive taxation between 1914 and 1980, the third factor in 'the great redistribution' was the liquidation of colonial assets and public debts. International investments had returned as much as 10 per cent of national income in Britain. 'The liquidation took place in two phases: the foreign assets were destroyed or transformed into public debts, and then the latter were themselves liquidated'.[36]

Consensus and Prosperity

The ideology of the period from 1945 to 1979 is often termed the Keynesian consensus.[37] But this puts an economistic gloss on what was a transformation of the whole of society. Judt uses the more correct term 'the social democratic moment'.[38] Hobsbawm calls the period 'the social revolution' (1945–1990).[39]

Though in Britain it was a period of economic growth, widespread prosperity, and greatly reduced inequality, there was also continuing industrial conflict as the powerful unions flexed their muscles. However, the broad institutions of the welfare state were maintained after Labour was defeated at the general election in 1951. The Conservative inheritors of the welfare state benefitted politically from what the Attlee government had forged, such that Conservative Prime Minister Harold Macmillan could remark to his Conservative Party colleagues in 1957, 'Let us be frank about it: most of our people have never had it so good'.

The pattern of transformation in Britain matched that observed in the USA and European countries. Piketty shows how inequality of income in the USA, Britain, Germany, France, and Sweden in varying degrees fell precipitiously between 1940 and 1950, then stabilised and rose again from around 1980. 'The top decile's share of total national income averaged around 50 per cent in Western Europe in 1900–1910 before falling to around 30 per cent in 1950–1980 (or even 25 per cent in Sweden)'.[40] Inequality of wealth fell even more dramatically: 'The top decile's share of total private property (real estate, professional, and financial assets, net of debt) was about 90 per cent in Western Europe in 1900–1910 before falling to around 50–55 per cent in 1980–1990, then rising again'.[41]

In the United States, the system of taxation became strongly progressive from 1914 to 1980. Piketty illustrates the change in a chart which shows that, 'the effective tax rates (all taxes combined, in per cent of total pre-tax income) paid by the highest incomes [richest 0.01, 0.1 and 1.0 per cent] were significantly higher than the average effective rate paid by the population as a whole'.[42] From about 1930 to 1950 they were greatly higher.

Annual figures for Britain's Gross Domestic Product (GDP) fluctuated regularly in the 25 years between 1949 and 1973 but averaged 3.5 per cent.[43] As Hall explains, the 'stop–go' pattern developed as a result of the policy conflict between the government's efforts to achieve full employment (resulting in inflation) and attempts to correct the balance of payments so as to avoid devaluation (of the pound sterling) resulting in deflation.[44] In the 37 years from 1982 to 2019, GDP growth averaged 2.4 per cent. In the 10 years, 2010 to 2019, the average annual growth was even lower at 2.0 per cent. In a series of charts, Piketty shows that growth of national income in the USA and Europe surged, with diminishing inequality and progressive taxation, and fell as inequality returned.[45]

Far from discouraging growth, the welfare state coupled with increased equality fed it. Piketty observes that strongly progressive taxation did not discourage innovation or productivity. He concludes,

> In sum, all the data at our disposal today suggest that virtually confiscatory tax rates have been an immense historical success. They have made it possible to greatly reduce the divergences of fortunes and incomes, while at the same time improving the situation of the middle and lower classes, developing the welfare state, and stimulating better economic and social performance overall[46]

The most enduring legacy of the welfare state in Britain, and the most widely supported, Judt argues, 'was not economic planning and state ownership but free health care, free public education, and subsidized transport',[47] and as it turned out, town and country planning with an emphasis on countryside protection rather than new town building. Conservative voters in the shires did not want to see 'their' countryside suburbanised.

Unfortunately, the forces that opposed social democracy, forming even as social democracy was being established, grew in strength, determined to return societies to the liberalism that held sway in the nineteenth century. When social democracy faced a crisis in the 1970s arising from the decline and international relocation of the traditional industries as well as external shocks, the ideology of 'neoliberalism' struck back in the 1980s, first of all under the political leadership of Margaret Thatcher.

Notes

1 Judt, T. (2005) p. 14.
2 In Chechnya, Russian forces almost completely destroyed the capital Grozny. During the South Ossetia–Georgia conflict, the European Court of Human Rights ruled that Russia was responsible for grave human rights abuses. Attacks on civilians and abusive acts by Russian armed forces in Ukraine are emerging on a daily basis.
3 National Archives, https://www.nationalarchives.gov.uk/cabinetpapers/themes/economy-second-world-war.htm (accessed 17/08/2022). See also Milward, A.S. (1947).
4 Milward, A.S. (1977).
5 Goodwin, D. (2001).

6 https://en.wikipedia.org/wiki/United_Kingdom_home_front_during_World_War_II (accessed 17/08/2022).

7 Judt, T. (2005) p. 76.

8 Ibid.

9 Addison, P. (1985).

10 Judt, T. (2005) p. 361.

11 Piketty, T. (2022) p. 145. Piketty compares that 'infamous tribute' to the same proportion of national income that France imposed on the rebellious colony of Saint Domingue (Haiti) in 1825 to compensate slave owners for the loss of their property.

12 Ibid. Chapter 6 'The "Great Redistribution" 1914–1980.

13 Piketty, T. (2020) p. 177.

14 Ibid. p. 175.

15 William Manchester, one of Churchill's biographers, called it a revolutionary concept because it was the first budget in British history with the expressed intent of redistributing wealth equally amongst the British population. Manchester, W. (1983).

16 Piketty, T. (2020) p. 180.

17 Gani, A. (2015) 'Clause IV, A brief history' *The Guardian* 09/08/2015). https://www.theguardian.com/politics/2015/aug/09/clause-iv-of-labour-party-constitution-what-is-all-the-fuss-about-reinstating-it (accessed 21/08/2022).

18 Bergstrom V. (1992) p. 142.

19 Ibid. p. 149.

20 Tilton, T. (1992) p. *410*.

21 Cited by Tilton (1992) p. 412.

22 Ibid.

23 Piketty, T. (2020) p. 186.

24 Quite a collection of laughably antique titles (where was the 'Lord High Executioner'?).

25 The office (of Lord Privy Seal) does not confer membership of the House of Lords, leading to Ernest Bevin's remark on holding this office that he was 'neither a Lord, nor a Privy, nor a Seal'. All the above is splendidly available on Wikipedia: https://en.wikipedia.org/wiki/Churchill_war_ministry (accessed 25/08/2022).

26 https://en.wikipedia.org/wiki/Beveridge_Report, (accessed 25/08/2022). But see also the report itself: *Social Insurance and Allied Services* held in the British Library.

27 The settlement movement, founded by Henrietta and Samuel Barnett, embodied the effort to get rich and poor people to live more closely together and work towards a future without poverty. Toynbee House was named after the historian Arnold Toynbee. There is something reminiscent of the pragmatic and perhaps paternalistic socialism of Robert Owen about the movement.

28 In *Unemployment: A Problem of Industry* (1909).

29 In *Planning Under Socialism* (1936).

30 Addison, P. (1985) p. 11.

31 Ibid.

32 Barlow Report (1940 edn.) p. 634.

33 'Location of Industry in Great Britain' (News) in *Nature* Vol 145. No. 3677, London: Macmillan.

34 Cullingworth, J.B. (1997) p. 18.

35 The Labour programme is usefully summarised in Wikipedia, (https://en.wikipedia.org/wiki/Attlee_ministry accessed 30/08/2022).

36 Piketty, T. (2022) p. 142.

37 See for instance Addison, P. (1985).

38 Judt, T. (2005) Chapter 11, p. 360.

39 Hobsbawm E. (1995) Chapter 10.

40 Piketty, T. (2020) p. 420, Figure 10.2.

41 Ibid. p. 423, Figure 10.4.

42 Piketty, T. (2022) Figure 22, p. 136.
43 https://www.statista.com/statistics/281734/gdp-growth-in-the-united-kingdom-uk (accessed 01/09/2022).
44 Hall, P.A. (1986) p. 50. Hall also points out that, 'As shown, British macroeconomic policy, both before and after the war [WW2], was dominated by the pursuit of a relatively high exchange rate and was consequently more deflationary than it might otherwise have been' (p. 51).
45 Piketty, T. (2020) pp. 544, 545, Figures 11.12, 11.13, 11.14, and 11.15. National income is defined as GDP minus capital depreciation, plus net income from overseas (p. 19 footnote 4).
46 Piketty, T. (2022) p. 139.
47 Judt, T. (2005) pp. 537-538.

References

Addison, P. (1985) *Now the War Is Over, A Social History of Britain 1945–51*, London: Jonathan Cape and the BBC.

Barlow Report. (1940 edn) *Report of the Royal Commission on the Distribution of the Industrial Population*, Command 6153, London: HMSO. See here: 'Location of industry in Great Britain' (News), *Nature* 145(3677).

Bergstrom, V. (1992) 'Party program and economic policy: The social democrats in government', in Misgeld, K., Molin, K., and Amark, K. eds. *Creating Social Democracy, A Century of the Social Democratic Labor Party in Sweden*, Pennsylvania: Pennsylvania State University Press.

Cullingworth, J.B. (1997) *Town and Country Planning in the UK*, London: Routledge.

Gani, A. (2015) 'Clause IV, A brief history', *The Guardian* (09/08/2015). https://www.theguardian.com/politics/2015/aug/09/clause-iv-of-labour-party-constitution-what-is-all-the-fuss-about-reinstating-it (accessed 21/08/2022).

Goodwin, D. (2001) 'The way we won: America's economic breakthrough during World War II', *The American Prospect*. https://prospect.org/health/way-won-america-s-economic-breakthrough-world-war-ii/ (accessed 17/08/2022).

Hall, P.A. (1986) *Governing the Economy, The Politics of State Intervention in Britain and France*, Cambridge: Polity Press.

Hobsbawm, E. (1995) *The Age of Extremes 1914–1991*, London: Abacus.

Judt, T. (2005) *Post War, A History of Europe Since 1945*, London: Penguin Books.

Manchester, W. (1983) *The Last Lion: Winston Spencer Churchill, Visions of Glory 1874–1932*. Boston: Little Brown.

Milward, A.S. (1977) *War, Economy and Society: 1939–1945*. Harmondsworth: Penguin.

Piketty, T. (2020) *Capital and Ideology*, Cambridge: Bellknap Press of Harvard University.

Piketty, T. (2022) *A Brief History of Equality*, Cambridge: Bellknap Press of Harvard University.

Tilton, T. (1992) 'The role of ideology in social democratic politics', in Misgeld, K., Molin, K., and Amark, K. eds. *Creating Social Democracy, A Century of the Social Democratic Labor Party in Sweden*, UniversityPark: ThePennsylvania State University Press, pp. 409–427.

4
THE NEOLIBERAL REGRESSION

Introduction

The neoliberal regression that began in earnest with the Thatcher government of Britain in 1979 transformed the world. It was based on a vision of society, environment, and economy from the mid-nineteenth century. This was a worldview of individual freedom, unconstrained markets, the virtue of property ownership, the dynamism of industry (then fed from colonies), and the salutary stimulus of poverty. At that period, a powerful state protected the rules of property entitlement and property transfer. A minimum of welfare was provided, just sufficient to guarantee subsistence.

Karl Polanyi believed that, with the Beveridge Report and the Attlee government, this nineteenth-century world view had been utterly transcended by the new vision and new institutions discussed in the previous chapter. Polanyi hoped and believed that human labour and the natural environment would henceforth be protected.

However, the global and technological context of the national economy was changing under the feet of governments. The welfare state had enriched a new middle class. The spirit of solidarity that marked the war years and post-war period had weakened. The wealth siphoned from the colonies was gone. Class conflicts between workers, owners, and the state resurfaced. Britain had neither the institutional apparatus nor the political culture to manage the emerging economy.

The neoliberal regression led not only to the privatization of nearly all public utilities but to the outsourcing of much government policy and planning work to consultants. Working both for private clients as well as governments, they do not provide unbiased advice (see Chapter 10). Weakening the capacity of the public service also weakens majoritarian democracy and inhibits economic redistribution – which is, of

DOI: 10.4324/9781003382133-5

course, exactly what Hayek wanted. Yet, contrary to Hayek, weakening the public service also places policymaking more directly on the shoulders of the politicians.

Instead of looking forward to the institutional changes that would equip the nation with the means to adapt to the new global context, the Thatcher government took the past as its model. In the end, the government could neither replicate the past nor prepare for the future.

Crisis

The crisis of the 1970s in Britain was nothing like the crisis that preceded the creation of the post-war welfare state. But it was significant, and the concatenation of events in the 1980s and 1990s flowing from the ideological shift back to market liberalism powered the ongoing structural change of 'globalisation'.

Initially, the crisis was one of adaptation of the British economy to the turn of events that Sir Montague Barlow had noted back in 1938: 'It may be expected that the importance of the old "basic" industries in the national economy may decline and that of other industries increase'.[1] As Judt remarks, 'The economic crisis, however circumstantial and conjunctural its triggers, coincided with a far-reaching transformation which governments could do little to arrest'.[2] But it was also a crisis of public expectations raised by the very success of the welfare state and the growing inability of national governments to meet those expectations.

By the 1970s, the world's production system was starting to undergo a massive change. On the one hand, human labour was being replaced by machines. On the other, manufacturing was moving away from the old industrial heartlands. Labour-intensive industries migrated from high-wage to low-wage countries such as China, South Korea, India, Mexico, and Brazil. The jobs lost would never come back.

Across Europe, national economies were further brought low by two external shocks. First, the post-war system of fixed exchange rates amongst national currencies was abandoned, initiated by President Nixon's unilateral decision to allow the value of the US dollar against other currencies to be determined by the international currency market – 'floating' the dollar. Once the value of the dollar against other currencies was to be determined by the foreign exchange market, so must those of other currencies. The floating currency system was confirmed at a conference in Paris in 1973.[3]

The second, the oil shocks, following the consequences of the attack on Israel by Egypt and Syria in 1973, provided further economic destabilisation and inflation. At a time when Western Europe was shifting away from solid fuel for energy and increasing its dependence on oil, the oil price more than doubled. Inflation increased from a steady 3.1 per cent between 1961 and 1969 to 11.9 per cent in the years 1973 to 1979. The world was plunged into recession.

In Britain, job losses accelerated in virtually every traditional industry. Judt records that even before 1973, the transformation was already under way in coal, iron, steel, engineering; thereafter it spread to chemicals, textiles, paper and consumer goods. Whole regions were traumatized: between 1973 and 1981 the British West Midlands, home of small engineering firms and car plants, lost one in four of its workforce.[4]

Between 1978 and 1986, unemployment grew from 5.5 to 11.3 per cent.[5] The proportion of long-term unemployment almost doubled.

Moreover, as political scientist Joel Krieger writes, in actions against Edward Heath's Industrial Relations Act: 'Between July, 1970 and July, 1974, more than 3 million work days were "lost" in strike actions ... and 1.6 million against the incomes policy (first voluntary, then statutory)'.[6]

Demographic changes further undermined the Keynesian economic assumptions. Baby boomers were retiring earlier, and the welfare state was expected to support them as they aged. Moreover, they were living longer, and birth rates were dropping. The result was ever-higher charges on income earners.

Judt insists that the 1970s collapse in Britain of the 'Keynesian' political consensus was not a result of ideological confrontation 'but a continuing failure of governments of all political colours to identify and impose a successful economic strategy'.[7] Political economist Peter Hall's analysis of the crisis is worth quoting at length here because of its echo in the global response to the Covid-19 pandemic and the fuel crisis generated by the war on Ukraine (massive stimulus, followed by inflation, followed by deflationary measures from central banks).

The oil embargo and massive oil price rises imposed by OPEC plunged the world into a prolonged recession. One nation after another attempted to revive its economy with a demand stimulus, only to find that unilateral expansion could not be sustained in an increasingly interdependent world. In medium-sized, open economies, such as Britain's, a stimulus brought inflation and an influx of imports that generated immediate balance of payments problems and an exchange rate crisis severe enough to force deflation again on the government.[8]

In the autumn of 1973, the Conservative government of Edward Heath tried to take on the National Union of Mineworkers. The Union demanded a 30 per cent wage increase. Heath refused and decided to outlast the strike that followed. The nation was put on a 3-day week and Heath called an election in February, in the depths of winter. He lost, and the incoming Labour government capitulated and met the mineworkers' demands.

During the next 4 years, the Labour government tried to reduce public expenditure. Despite an attempt by the unions and government to agree on a prices and incomes policy, the government could offer the unions no real benefit in exchange for wage restraint. There followed the 'winter of discontent' in 1978–79. Several public sector unions defied the incomes policy. 'For a period, they shut down the hospitals, graveyards and garbage collections across England'.[9]

The British Labour Party was unwilling to employ interventionist policy to increase the profitability of the private sector. 'Therefore', as Hall argues,

> the British electorate has never been offered the option taken up by the French or Japanese: a policy whereby the state employs an extensive array of sanctions and incentives to enforce modernization on industry so as to favour its most profitable and growing segments.[10]

Nor did Labour seriously consider the Scandinavian and West German models of industrial relations, including workers' representatives on the boards of companies, which might have avoided the industrial confrontations that beset Britain in the 1970s and early 1980s.

It turned out that Keynesian policies in a globalised world required the consent of working people to tailor their demands for higher wages to the need to avoid inflation. After 1975, the Labour government tried to establish a prices and incomes accord with the trade unions (through the Trades Union Congress), but the need to cut public expenditure meant that the government had little to offer the unions as a trade-off for wage restraint. A reasoned consensus between major firms, state agencies, and labour proved unachievable. The Labour Party's paymasters, the industrial unions 'preferred nineteenth-century style confrontation – which they stood a good chance of winning to negotiated – contracts signed in Downing Street that would bind their hands for years ahead'.[11]

As Hall points out, citing Samuel Beer, 'Political parties are the agents of collective purpose in a democracy. As such they distil the multiple strands of social sentiment into concrete programmes backed up by a particular moral vision'.[12] Policies build on an accumulation of ideas, forged by the party into electoral appeal. This seems to be the reason why national policymaking so often fails to absorb lessons from policies developed outside national boundaries, giving meaning to the adage, 'all politics is local'.

By the end of the 1970s, the British public had had enough of conflict between the trade unions and the government. By 1979, many began to view inflation as a bigger threat than unemployment. Seventy-nine per cent of the electorate and even 50 per cent of unionists themselves believed that the trade unions had too much power.[13]

Turning again to Judt, 'If the European state could no longer square the circle of full employment, high real wages and economic growth, then it was bound to face the wrath of those constituents who felt betrayed'. In particular, 'It was the heavily taxed middle classes – white collar public and private employees, small tradesmen and the self-employed – whose troubles translated most effectively into political opposition'.[14]

Critics of the social democratic form of government argued that the state should be distanced from the economy, upholding simple rules of the market. The state should not own or intervene in the means of production. Most social services provided by the state could more efficiently be provided by the private sector. Such ideas were canvassed at a conference of the Conservative Party at Selsdon Park, bringing down thunderous criticism from the political establishment nurtured on Keynesian orthodoxy. Yet even the Labour Party under PM James Callaghan had, following the 'winter of discontent', begun to deviate from the mantra of full employment, seeking to bring down inflation at the cost of unemployment, economic hardship, and slower growth.[15]

Ideology

The ideology of what became known as 'neoliberalism' was founded on the work of critics of social democracy stretching back to pre-war days and even earlier. In fact, the ideology should more properly be called the neoliberal regression. It was not fundamentally new. Its neo-ness referred to the way classical liberalism had now to accommodate the social protections wrought by social democracy and the growing power of the middle class. But in essence it sought a return to a 'proprietarian' society. That is to say, in Piketty's words, 'A social order based on a quasi-religious defence of property rights as the *sine qua non* of social and political stability'.[16]

The living link between the liberalism of the nineteenth century and the liberalism in force today was the Austrian economist and political philosopher Ludwig von Mises, who fled Austria for the USA in 1940. In his long life (1881–1973) Mises expounded the philosophy and ideology of classical liberalism: 'The program of liberalism', he wrote, 'if condensed into a single word, would have to read: property, that is, private ownership of the means of production'.[17]

Polanyi's critique of liberalism (discussed in Chapter 2) was directed at the naturalistic fallacy, held to be true by Mises. This ideology posited that the division of labour was an inherent fact of human behaviour, and that the realistic self-protection of human society from the institution of the market ('protectionism') was 'due to impatience, greed and short-sightedness, but for which the market would have resolved all its difficulties'.[18] Mises's work anticipated that liberalism must be global in scope. National sovereignty, he said, was an illusion. 'The nations', he argued,

> must come to realize that the most important problem of foreign policy is the establishment of lasting peace, and they must understand that this can be assured throughout the world only if the field of activity permitted to the state is limited to the narrowest range.[19]

Mises, with his disciple, Friedrich von Hayek, founded the Mont Pelèrin society in 1947 to promote the return of liberalism and rebuild it as the world-wide hegemonic ideology. Hayek drew on the example of the spread of socialism to guide his political strategy:

> In every country that has moved towards socialism, the phase of the development in which socialism becomes a determining influence has been preceded for many years by a period during which socialist ideals governed the thinking of the more active intellectuals[20]

The Mont Pelèrin Society 'was committed to persuading the intellectuals, and hence the masses and their political leaders to change course'.[21]

Like Mises and the classical theorists, Hayek formulated the idea of liberalism from a set of assumptions about the 'rationality' of human behaviour, rather than from any empirical historical observation of society. He followed Mises in 'praxeology' – a theory of human action, based on the notion that humans engage in rational behaviour. Thus, liberals construct their ideology from their beliefs and not from facts.

Hayek's subtlety was, however, to realise that liberal-leaning intellectuals constituted a broad church, with variable understandings of the practice of liberalism in the modern world. The Mont Pelèrin Society brought together a range of ideas about liberal praxis, ranging from the ordo-liberalism of the German group to the Chicago school espousing monetarism, and public choice theory applying rationalism to political behaviour.[22]

The liberalism that was inflicted upon Britain from 1981 was a mixture of Hayek's ideas about the role of the state, the monetarist theory of Milton and Rose Friedman,[23] the Chicago school, and some liberal political science ideas such as public choice theory and 'political overload'. Public choice theory proposed that the assumptions about individual behaviour from liberal economics should also be applied to political decision-making, namely that society is nothing more than the aggregate of individual behaviour, and individual behaviour including political behaviour is directed at material self-advantage. The concept of political or government overload held that the British state had grown too large, displacing activity that could be performed more efficiently by the private sector.[24]

Hayek feared any form of redistribution of income and wealth, especially progressive taxation. He proposed to prevent such a contingency by severely restricting democracy. 'Law' he defined as natural law, not laws made by governments. In the final volume of *Law, Legislation and Liberty* he proposed that the sacred law of property should be protected by a supreme assembly consisting of persons of 45 years of age or older chosen to serve 15-year terms after having demonstrated their abilities and professional success.[25] Hayek is reported as saying in 1981, 'Personally, I prefer a liberal dictatorship to a democratic government without liberalism'.[26]

The dictatorial Pinochet regime in Chile acted literally on Hayek's words. No such option was available to the Thatcher government without overturning the

British Constitution. However, Thatcher espoused strong government to reinstate liberalism, thereby creating a contradiction: Hayek's preference was for enforcement of liberalism *outside* politics, Thatcher (and Pinochet, for that matter) enforced liberalism *within* politics. That option left open the possibility of clientelism: political decision-making being captured by particular business interests, just as Labour governments had been (so Conservatives argued) captured by trade union interests. Hayek himself, however, regarded capture of government by particular business interests as the worst possible compromise.

The one ideology Thatcher embraced from the start was monetarism. The theory proposed that the rate of inflation could be controlled by maintaining low growth of the money supply. Government discretionary decisions on regulating demand for money (through wages) should be replaced by simple rules on supply that dictated the appropriate level of monetary growth. 'If these rules were apparent to all, they would warn the unions off excessive pay settlements because, with restricted monetary growth, those would only result in rising unemployment'.[27]

Monetarism is a version of the belief in self-regulating markets. It found support among City of London economists who found the theory useful for understanding the relationship between monetary aggregates as well as supportive of their preference for firmly anti-inflationary measures. It remains the weapon of choice of central bankers today: inflation must be controlled by reducing the money supply, raising interest rates, and reining in access to credit, thereby creating unemployment. Monetarism triumphed, as Hall argued,

> not as economic science, but as political ideology. Its validity could not be proven in scientific terms, but to those who were seeking some way to restore the authority of the state it seemed to offer a solution ... It was also a doctrine designed to justify an attack on the power of the unions.[28]

The institutionalist economist Geoffrey Hodgson, writing of Thatcherism just a few years after her election, considered monetary theory to be a 'smokescreen' providing an intellectual justification for a more profound ideological change. He cites Thatcher's revealing response to a question from a *Guardian* journalist in 1981: When asked in February 1981 whether unemployment in excess of 3 million might lead her to decide that the human cost of monetarist theory was too high, Thatcher replied: 'It is not a theory. It is borne out by everything that has happened in this country in the last thirty years'. The 'last thirty years' were the years of social democracy. Hodgson comments: 'It was not simply a matter of "abandoning" Keynesian *theories*, but of rejecting Keynesian and social democratic *objectives*: government guidance and a publicly financed welfare state'.[29]

Political Champions

When Margaret Thatcher won power for the Conservatives in 1979, her famous TINA statement, 'There Is No Alternative' was only partly true. There *was* a social-democratic alternative for industrial relations. But ideological alternatives could not just be picked from a menu. They are formed, as they were in Scandinavia, Germany, Austria, and Australia, from particular national political and social histories.

Like the Attlee government, the Thatcher government was preceded by intellectual champions. First among them were Mises, Hayek, and the members of the Mont Pelèrin Society. Sir Keith Joseph was influenced by the ideas promoted by the Society. He was a barrister, company director, and Member of Parliament from 1956 and held ministerial posts in Conservative governments from those of Harold Macmillan to Edward Heath. He became Thatcher's most influential advisor.

Joseph favoured the principle of the 'social market economy', a concept derived from the 'ordo-liberal' branch of neoliberalism led by Wilhelm Röpke. The ordo-liberals did not believe that the market alone would create a humane society. They feared the concentration of power in monopolies and cartels as much as in the state. Economic efficiency had to be accompanied by humanitarian principles in a 'community of people'.[30] The idea of the social market economy was further developed by Alfred Müller-Armack to influence the conduct of Germany's economy in the 1960s.[31]

But the social protections institutionalised in Germany, such as worker participation on company boards, were not transplanted into Britain under the government of Thatcher. At a time of industrial strife and with a tradition of union confrontation with the state, Thatcher preferred to take on and defeat the unions. Instead, Thatcher joined Joseph in adopting the ideology of monetarism. They, together, set up a 'think tank' – the Centre for Policy Studies (CPS) – to develop advice on free market policies. The CPS describes itself today as 'Britain's leading centre–right think tank. Our mission is to develop a new generation of conservative thinking, built around promoting enterprise, ownership and prosperity'.[32]

Similar neoliberal or neoconservative 'centres' proliferated around the world urging a return to the ideas of nineteenth-century liberals and conservatives. Judt comments that the free market principles of neoliberalism were no more than,

> simple nostrums of pre-Keynesian liberals brought up on the free market doctrines of neoclassical economics ... since 1973, however, free-market theorists had re-emerged, vociferous and confident, to blame endemic economic recession and attendant woes upon 'big government' and the dead hand of taxation and planning that it placed upon national energies and initiative[33]

Margaret Thatcher herself was the daughter of Methodist parents who ran a grocer's shop. She was always a Conservative, won a scholarship to Oxford and studied chemistry. She rose to be president of Oxford's Conservative Society, but her education in science later showed itself in her acceptance of the science of climate change.

She was not liked by the old guard of the Conservative Party, and not much by the voting public. Judt observes, 'Under Mrs Thatcher's leadership the Conservative Party never actually gained many votes. It did not so much win elections as watch Labour lose them, many Labour voters switching to Liberal candidates or else abstaining altogether'.[34] In this way, the Thatcher transformation was the opposite in many ways of the social democratic transformation that brought Attlee to power. There was no public consensus around neoliberalism. Nobody outside the business community and a few ideologues clamoured for the policies Thatcher delivered: reduced taxes, privatization of industries and services, 'rolling back the state', and 'family values'.

Thatcher consolidated power around herself and her government, enabling the central state to enforce her policies. In administration, Judt notes, 'she was an instinctive centralizer'. Government ministries found themselves constrained by the Prime Minister and a small coterie of friends and advisors.[35]

Consequences

The consequences of Thatcher's transformation spun out from relatively modest beginnings to become the neoliberal orthodoxy we live with today. It is worth tracing first the more immediate institutional shifts in Britain before considering the wider impact of the neoliberal turn.

Thatcher was driven by political instinct and a belief in strong, undeviating government as much as by strict application of economic ideology. She drew support both from those of her colleagues who did not shrink from intervention in the economy and those who believed in *'laissez faire'* – the attitude that if left alone markets would be self-regulating. Thatcher created a hybrid institutional regime that combined strong political intervention with a laissez faire approach to the productive economy, an approach that became characteristic of neoliberalism internationally.

Her early goals on forming a government were to limit the growth of the money supply, reduce the public sector deficit, cut public spending, and reduce taxation. The monetarist macroeconomic strategy was not easy to implement. Social protections of the welfare state embedded under social democracy could be reduced, but not reversed. Despite her attack on big government, Thatcher was unwilling to dismantle any of the major programmes associated with the welfare state in Britain. Public opinion polls showed that 'even avid critics of government inefficiency did not want to see substantial cuts in the National Public Health Service, primary education, unemployment insurance or old age pensions'.[36]

The electoral platforms of both the Thatcher government in Britain and the Reagan presidency in the United States argued for monetary restraint and cuts in public expenditure and taxation. But while Reagan made cuts in taxation his priority, he financed spending on social protection and the military by public sector borrowing, Thatcher chose to forgo tax cuts in order to reduce the public sector deficit.[37]

While America under Reagan, therefore, experienced a Keynesian stimulus, Britain, by the end of 1980, went into severe recession. Unemployment doubled between 1979 and 1981. GDP fell by 2.2 per cent in 1980 and 1.1 per cent the following year. The rapid rise in the rate of exchange of the pound on currency markets undermined the international competitiveness of British industry.[38] Strict monetary policy had the adverse effect of forcing the economy into recession, destroying manufacturing industry in particular, and burdening the national insurance system with ever-increasing unemployment payments.

The fear of unemployment rose as the unemployment rate surged to over 4 million in 1984.[39] This unintended consequence of monetarism served Thatcher's wider ideological purpose well. However, yet greater political force, Thatcher judged, was required. The government shifted towards a harsh industrial relations regime. Long battles were fought with the unions in the 1980s culminating in the draconian crushing of the National Union of Mineworkers' strike in 1984, with bitter clashes between striking miners and dragoons of police brought in from all over Britain.

A series of Acts of Parliament were passed limiting the power of the unions: abolishing the right to picket beyond an employee's own place of work, making it difficult to secure 'closed shops' (which required union membership for anyone to be employed in a firm), limiting the right to appeal against unfair dismissal, requiring majority union member approval before holding a strike, and imposing large fines for breach of any of these laws.

Thatcher's industrial policy was geared to the return to private ownership of major nationalised industries. By 1985 'at least 50 per cent of the shares in over a dozen companies formerly under state ownership had been sold to the private sector'.[40] The list included many nationalised enterprises (such as British Aerospace, the Jaguar car company, and British Rail Hotels) that properly belonged in the competitive arena of the private sector. Others, however, provided public services (for example, British Gas, British Telecom, and Thames Water).

Without industrial privatisations, which raised over £7 billion, the government would have had to raise taxes or make spending cuts on social services. Beyond pragmatic politics there was also an ideological purpose to these measures. As Margaret Thatcher said, 'We need to create a mood where it is everywhere thought morally right for as many people as possible to acquire capital'.[41]

However, if the aim was to spread capital ownership and increase competition, in Hall's view, privatization did little to achieve it. Many of those who bought shares in privatised firms went on to sell them to large institutional investors. Public service corporations in fields such as water, electricity, gas supply, and public transport held natural monopolies, allowing the newly privatised firms in these fields to exploit their monopoly position for private gain. Hall observes, 'Although the activities of such firms will be regulated, it is not clear that regulation is any more efficient in these areas than nationalization'.[42]

The government's aim, in any case, was to loosen regulation of the private sector. The requirement for large firms to report statistical material was reduced. Small businesses were relieved of many planning restrictions. Twenty-four urban areas were designated 'enterprise zones' in which planning restrictions were lifted and tax relief granted to encourage firms to locate there. Large amounts of money were channelled into unprofitable public enterprises such as British Leyland and the British Steel Corporation, not with the aim of supporting 'lame ducks' but to keep them afloat while making them profitable at any cost (in lost employment or forgone capital investment). The ultimate aim was to make these firms attractive for privatization.

The most enduring social legacy of the neoliberal regression was the redistribution of real income and wealth resulting in the growth of inequality and poverty. As noted earlier, in the late 1970s, structural factors in the global economy ended 30 years of economic prosperity: the onset of globalisation eroding manufacturing employment and the oil shocks raising the costs of production.

Hall contrasts the response of the Thatcher government to these structural shifts with the preceding Labour government's response. Both Labour and Conservative governments of the 1970s had to cope with rising unemployment, inflation, and slower economic growth. To meet these challenges both the Thatcher government and its Labour predecessor instituted austere economic policies with distributional effects across classes, regions, and socioeconomic groups. The Labour government of the seventies tried to mitigate distributional effects and organise the distribution of income through tripartite negotiations involving government with trade unions and business representatives (softening the impact of austerity on groups dependent on state transfers such as unemployment and pension benefits).

The Thatcher government, following neoliberal ideology, had no such intention to mitigate distributional outcomes. The government based its policies on 'a strategy of reinforcing market mechanisms and a market-led allocation of resources so as to increase work incentives and the flexibility of factor inputs [capital and labour: fictional commodities in Polanyi's terms] in the economy'.[43]

Economic recession and rising unemployment tended to reinforce existing inequalities in income and labour market power. But the Thatcher government increased these inequalities by deliberate strategy. The effects were felt in three ways: through high levels of unemployment, changes in transfer payments and taxation, and housing policy.

The principle of full employment governing macroeconomic policy was changed to one of fighting inflation first and tolerating the resulting unemployment, even at high levels. The most obvious cost of the Thatcher strategy 'has been a massive increase in the number of jobless and the poor'.[44] The result was a nation divided on regional, racial, and age-related lines. By 1984, the unemployment rate in the South of England had risen to between 8 per cent and 9 per cent as against 15 per cent in Scotland, Wales, and Northern England. Wages were 20 per cent higher in the South East than wages in the North. 'Whole communities suffer from the effects of mass unemployment in their area'.[45] Unemployment reached 20 per cent among the under-20 age group and 50 per cent among black youth in the inner cities. Overall, those with jobs did well enough 'but the unemployed and the poor bore the brunt of the cost of industrial adjustment and the fight against inflation'.[46]

Taxation policy was recast to shift the tax burden from incomes to consumption (Value Added Tax). The former remained progressively structured (with those on higher incomes paying a larger proportion of their income in tax than those on lower incomes), but consumption or expenditure taxes were simply proportional to expenditure on products and services purchased. Moreover, 'upper-income earners continued to benefit from the large (tax) reductions they received in the government's first budget, while lower-income earners found that increasing portions of their transfer payments were taxed'.[47]

The Thatcher government reduced the real value of transfer payments. Old age pensions fell slightly in real terms. The value of child support for an unemployed couple fell by 20 per cent. The government broke the index-linked (to the CPI) value of unemployment benefits, abolished the earnings-related supplement and subjected the benefits to tax. 'Although difficult to document', Hall suggests, 'Cuts in ancillary services provided by local social workers, councils, and the health service may have had an even more dramatic impact on those living below the poverty line'.[48]

The government made the sale of council housing and reduction in the public housing stock a central part of its social programme. Funding for construction, repair, and subsidy of public housing was cut by 55 per cent between 1979–80 and 1984–85. Dependence of low-income families on public housing meant that these cuts 'had a major impact on disparities between the affluent and the less affluent'.[49]

Turning to the longer-term perspective on Margaret Thatcher's neoliberal transformation, Judt makes an assessment from a historian's viewpoint some 25 years after she gained power. According to Judt, Britain's economic performance improved after the initial recessions. Inefficient firms were 'shaken out', the unions were 'muffled', and productivity and business profits rose sharply.[50]

Others, from an economic perspective, contest these claims. Cribb, Hood, and Joyce (writing in 2016) showed that incomes stagnated across much of the British economy since the 1980s.[51] In their 40-year assessment of Thatcher's legacy, Albertson and Stepney (writing in 2020) argue that, 'The sea change which occurred in UK economic prospects in the 1980s was not a change for the better; per capita real income growth slowed markedly post-1979. Since 1979, each government has underperformed its predecessor in this regard'.[52] These authors also argue that there is no evidence that privatisation led to efficiency gains which benefited the British people.

Paul Hirst, a sociologist (writing in 1989), was therefore quite accurate in his assessment of Thatcherism as a mirage. Hirst points out that Thatcher, by simply disempowering the unions, made no attempt to restructure industrial relations away from the conflictual model. The possibility of collaboration between workers and employers at Board level – as proved successful in Germany and Scandinavia – was not considered. Nor did the government attempt to create cooperative networks of industrial relationships between subcontractors and large firms as found in Japan and South Korea.[53]

Judt is more critical of Thatcher's social legacy. As a society Britain, he writes, suffered meltdown, with catastrophic long-term consequences:

> By dismantling all collectively-held resources, by vociferously insisting upon an individualist ethic that discounted any unquantifiable assets, Margaret Thatcher did serious harm to the fabric of British public life ... with everything from bus companies to electric supply [and the railway service] in the hands of competing private companies, the public space became a market place.[54]

Thatcher's policies amounted to 'class warfare which was the very stuff of politics'.[55] She faced Britain with a stark choice: between doctrinaire socialism and capitalism. She won the public to her cause not because they liked the cause but because the Labour Opposition could not offer an acceptable alternative. Labour was mired in the memory of industrial conflict that challenged the stability of the state. What mattered most to British voters, Judt argues, 'was not economic planning or state ownership but free medicine, free public education and subsidised public transport'.[56] These facilities were not very good, Judt continues, 'the cost of running a welfare state was actually lower than elsewhere, thanks to under-funded services: inadequate public pensions, and poor housing provision – but they were widely perceived as an entitlement'.[57]

Joel Krieger argues that the swing of voting from Labour to Conservative in 1979 among skilled and unskilled manual workers was not so much a rejection of the welfare state as:

> A call by each and every one for a greater share in the good life – the unfulfilled promises of the welfare state. Whether the provisions were located in the public or private sphere mattered rather little, but someone should provide good, clean affordable housing, houses that felt like homes, that buoyed the spirits. Social democracy in its shabbiness and torpor failed to replace pride in ownership with pride in community.[58]

Thatcherism finally lost its popular support after the government proposed to replace council rates based on the value of property with a poll tax applying at the same rate to all properties. New Labour under Tony Blair (from 1997), rather than rethink social democracy for new conditions, simply accepted the neoliberal regression. Judt writes, 'Although New Labour remained vaguely committed to "society", its Blairite leadership group was as viscerally suspicious of "the state" as the most doctrinaire of Thatcherites'.[59]

The 'great redistribution' precipitated by the two World Wars and brought to a head by governments after the Second World War caused the inequalities of pre-war 'ownership society' to 'fall off a cliff' (in Piketty's words). But inequality of income and wealth began to grow again in most parts of the world from 1980 'whether it be social-democratic Europe, the United States, India or China ... with a strong rise in the top decile's [top ten per cent] share of total income and a significant drop in the share of the bottom 50 per cent'.[60] Inequality increased least in the social-democratic societies of Europe which offered greater social protection than the neoliberal model embraced by Thatcher and Reagan.

Despite the deficiencies of monetarist policies, monetarism became a central element of the Washington Consensus. Central banks shifted their priorities from guaranteeing full employment to fighting inflation. That all changed after the global financial crisis when governments around the world borrowed heavily in order to stem the worst effects of the 'Great Recession' (from 2008). The Covid-19 pandemic again turbocharged the temporary return of Keynesian policies, flooding economies with cheap money.

Piketty comments: 'The spectacular increase in monetary creation since 2008 illustrates once again to what point economic institutions are not unchanging. They are constantly redefined according to crises and power relationships, within unstable and precarious compromises'.[61] So, as Piketty continues, there is no good reason not to increase the money supply if it enables us to finance useful policies such as the struggle for full employment, a guaranteed job, the thermal insulation of buildings, or public investments in health care, education, and renewable energy. Inversely, if inflation flares up in the long term, then that means that the limits of monetary creation have been reached and that it is time to rely on other tools to mobilize resources (beginning with taxes).[62]

Those limits were reached in 2022, but the critical lesson about the limits of monetarism have not been learnt. Following the Ukraine War and the after-effects of the Covid-19 crisis on supply chains, inflation began to surge. But it was inflation on the supply side: *price* inflation, not *wage* inflation. That has had a huge effect on housing markets and the distribution of wealth to existing homeowners and investors in housing and land. In response to government efforts to stop the spread of the virus through lockdowns that brought business to a halt, societies were flooded with money, which, with limited supply of housing, caused the price of housing to rise dramatically. Costs rose for new building (on the supply side), but the main effect was felt on the price of land, and thus on the price of existing housing.

The long-term impacts of the neoliberal regression are still playing out. As Robert Cox has argued, 'economic multilateralism meant the structure of world economy most conducive to capital expansion on a world scale; and political multilateralism meant the institutionalized arrangements made at that time and in those conditions for inter-state cooperation'.[63] There is a transnational process of consensus formation among the official caretakers of the global economy.

This consensus is shaped in meetings such as the World Economic Forum, the Trilateral Commission, and the Bildeberg Meetings, as well as global bodies such as the Organization for Economic Cooperation and Development (OECD), the International Monetary Fund (IMF), and the World Trade Organisation. Camilleri points out:

> The most powerful ministries and agencies of government are those that are most closely connected with the global economy and its institutions. This may help to explain why, despite wide national variations in political traditions and institutions, so many advanced capitalist states have, since the late 1970s, abandoned the strategy of Keynesian demand management and substantially lowered the expectations associated with the welfare state.[64]

Free trade has never been simply about allowing countries, without institutional barriers such as tariffs, to export their products on world markets and import what they can't produce from other countries. International trade is also about investment.

Global investment has meant free movement of capital, and power disparities between rich and poor countries. The free circulation of capital has provided the owners of capital with a major new source of power. As Piketty argues:

> The challenge to the welfare state and progressive taxation since the 1980s was not based solely on talk. It was also materialised in a set of rules and international treaties seeking to make a change as irreversible as possible. The heart of the new rules is the free circulation of capital, without any compensation in the form of regulation or common taxation.[65]

Free movement of capital necessarily demands free movement of labour, as Polanyi explained. The institutional creation of a *national* economy necessitated freeing workers to move beyond their parishes to find work. Today, in the *global* economy, labour remains tied to national boundaries. The European Union is the only transnational institution where free movement of labour across national boundaries is a foundational principle. Even in the Union, free movement of labour has generated tensions and nationalist sentiments resulting eventually in the withdrawal of Britain. Elsewhere, restrictions on 'immigration' remain severe. Refugees from persecution may be tolerated, but international jobseekers are demonised as 'economic refugees' or worse.

The structural contradiction between global investment and labour tied to nations remains unresolved. The tension between social protection and the market economy that Polanyi identified is today producing a rise in nativism: the protection of national identities and cultures. In Britain, nativist ideology triumphed over liberalism to deliver Brexit and six years of so-called 'populist' government (from Johnson to Truss and Sunak), in which class warfare continued under the nativist cloak. The result was a series of increasingly farcical, corrupt, and incompetent governments.

The discussion of these issues, however, takes us far beyond the scope of this chapter, focused as it is on the neoliberal transformation in Britain. That wider discussion will be deferred to Part 2 of the book examining the present crisis, the future society, and the climate emergency. Before opening up the discussion further, however, we need to turn to the next case study of transformation: from communism to capitalism in Russia.

Notes

1 Barlow Report (1940 edn.) See here: 'Location of Industry in Great Britain' (News) in *Nature* Vol 145. No. 3677, London: Macmillan. P. 634.
2 Judt, T. (2005) p. 458.
3 Ibid.p. 454. The British pound was devalued in 1972. In March 1973 the German Bundesbank Central Council decided to abandon the 'Bretton Woods' system of fixed exchange rates which had prevailed since 1944. (https://www.bundesbank.de/en/tasks/topics/1973-the-end-of-bretton-woods-when-exchange-rates-learned-to-float-666280).
4 Judt, T. (2005) p. 459
5 https://www.ons.gov.uk/employmentandlabourmarket/peoplenotinwork/unemployment/timeseries/mgsx/lms (series 290623xls): unemployed between age 16 years and over seasonally adjusted (accessed 29/06/2023)
6 Krieger, J. (1986) pp. 69-70.
7 Judt, T. (2005), p. 538.
8 Hall, P.A. (1986) p. 94. (Peter A. Hall is Krupp Foundation Professor of European Studies at Harvard University).
9 Ibid. p. 95.
10 Ibid. p. 92.
11 Judt, T. (2005) p. 538
12 Hall, P.A. (1986) p. 91, citing Beer, S.H. (1969) *Modern British Politics*, London: Faber.
13 Ibid. p. 95.
14 Judt, T. (2005) p. 462.
15 Ibid. p. 538-539.
16 Piketty, T. (2020) Glossary: p. 1044.
17 Mises, L. von (1985) p. 18.
18 Polanyi, K. (1944) p. 148.
19 Mises, L. von (1985) p. 144.
20 Hayek, F. (2005) p. 106.
21 Turner, R.S. (2008) p. 71.
22 See Low, N. (2020) pp. 88-95. Ordoliberalism, it should be noted, informed the ideology of the European Union of which Britain became a member in 1973: 'a doctrine according to which the essential role of the state is to guarantee the conditions of "free and undistorted" competition'. (Piketty, 2020: p. 706).
23 Friedman M. with Friedman R.D. (1962).
24 See Hall, P.A. (1986) p. 100, and Mueller, D.C. (1989) The range of neoliberal ideology is addressed more fully in Low, N.P. (2020) pp. 84-98.
25 Hayek, F. (1979).
26 Cited by Piketty, 2020: p. 709 footnote.
27 Hall, P.A. (1986) p. 96.

Hall also quotes Sir Geoffrey Howe, the Government's first Chancellor of the Exchequer [treasurer]: 'If workers and their representatives take pay decisions which are unwise because they seek too much, they will find they have crippled their employers and gravely harmed themselves by destroying their own jobs' (Ibid. p. 101).

28 Hall, P.A. (1986) p. 99.
29 Hodgson, G. (1984) p. 189.
30 Turner, R.S. (2008) p. 83.
31 Müller-Armack, A. (1965).
32 https://cps.org.uk/about/ (accessed 18/10/2022)
33 Judt, T. (2005) p. 537.
34 Ibid. p. 541.
35 Ibid.
36 Hall, P.A. p. 115.
37 Ibid. p. 103.
38 Ibid. p. 105.
39 The Permanent Secretary to the Treasury remarked, 'What has emerged in shop-floor behaviour through fear and anxiety is much greater than I think could have been achieved by more cooperative methods' (*The Times* London, 31/03/1983, cited by Hall, P.A.(1986) p. 109.
40 Hall, P.A. (1986) p. 110.
41 Ibid. p. 111 (quoted in *The Economist*, 19/10/1985 p. 70).
42 Ibid. p. 111.
43 Ibid. p. 123.
44 Ibid. p. 125.
45 Ibid. p. 125.
46 Ibid. p. 125.
47 Ibid. p. 126.
48 Ibid.p. 125.
49 Ibid. p.125.
50 Judt, T. (2005) p. 542.
51 Cribb, J., Hood, A. and Joyce, R. (2016).
52 Albertson, K. and Stepney, P. (2020) p. 321.
53 Hirst, P. (1989).
54 Judt, T. (2005) p. 543.
55 Judt, T. (2005) p. 545.
56 Ibid. p. 537.
57 Ibid. p. 538.
58 Krieger, J. (1986) pp. 85-86.
59 Judt, T. (2005) p. 547.
60 Piketty, T. (2020) p. 491.
61 Piketty, T. (2022) p. 240.
62 Ibid.
63 Cox, R.W. (1992) p. 162.
64 Camilleri, J.A. (1995) p. 214.
65 Piketty, T. (2022) p. 170.

References

Albertson, K. and Stepney, P. (2020) '1979 and all that: A 40-year reassessment of Margaret Thatcher's legacy on her own terms', *Cambridge Journal of Economics*, 44, 319–342.

Barlow Report. (1940 edn) *Report of the Royal Commission on the Distribution of the Industrial Population*, Command 6153, London: HMSO.

Beer, S.H. (1969) *Modern British Politics*, London: Faber.

Camilleri, J.A. (1995) 'State, civil society, and economy', in Camilleri, J., Jarvis, A.P. and Paolini, A.J. eds. *The State in Transition, Reimagining Political Space*, Boulder, CO: Lynne Rienner Publishers.

Cox, R.W. (1992) 'Multilateralism and world order', *Review of International Studies*, 18(2), 161–180.

Cribb, J., Hood, A., and Joyce, R. (2016) 'The economic circumstances of different generations: The latest picture', IFS Briefing Note BN187, Institute for Fiscal Studies. https:// www.ifs.org.uk/uploads/publications/bns/bn187.pdf (accessed 27/09/2022).

Friedman, M. withFriedman, R.D. (1962) *Capitalism and Freedom*, Chicago: University of Chicago Press.

Hall, P.A. (1986) *Governing the Economy, the Politics of State Intervention in Britain and France*, Cambridge: Polity Press.

Hayek, F. (1979) *The Political Order of a Free People: A New Statement of the Liberal Principles of Justice and Political Economy*, London: Routledge and Kegan Paul

Hayek, F. (2005) *The Road to Serfdom with The Intellectuals and Socialism*, London: Institute of Economic Affairs.

Hirst, P. (1989) *After Thatcher*, London: Collins

Hodgson, G. (1984) 'Thatcherism, the miracle that never happened', in Nell, E.J. ed. *Free Market Conservatism, A Critique of Theory and Practice*, pp. 184–208, London: George Allen and Unwin.

Judt, T. (2005) *Post War, A History of Europe Since 1945*, London: Penguin Books.

Krieger, J. (1986) *Reagan, Thatcher and the Politics of Decline*, Cambridge: Polity Press.

Low, N.P. (2020) *Being a Planner in Society, for People, Planet, Place*, London: Edward Elgar.

Mises, L. von (1985, first published 1962) *Liberalism in the Classical Tradition*, San Francisco: The Foundation for Economic Education; Cobden Press.

Mueller, D.C. (1989) *Public Choice*, Cambridge: Cambridge University Press.

Müller-Armack, A. (1965) 'The principles of the social market economy', *The German Economic Review*, 3(2), 87–114.

Piketty, T. (2020) *Capital and Ideology*, Cambridge: Belknap Press of Harvard University.

Piketty, T. (2022) *A Brief History of Ideology*, Cambridge: Belknap Press of Harvard University

Polanyi, K. (1944, 2001 edition) *The Great Transformation, The Political and Economic Origins of Our Time*, Boston: Beacon Press.

Turner, R.S. (2008) *Neo-liberal Ideology, History, Concepts and Politics*, Edinburgh: Edinburgh University Press.

5

COMMUNISM TO CAPITALISM

Introduction

In the late 1980s, as state communism was collapsing under the weight of its internal contradictions, neoliberalism was resurgent internationally. By 1989, the concept of the 'Washington Consensus' was formed to describe the position of Washington-based institutions of global aid and governance such as the International Monetary Fund and the World Bank. Initially, the target for these institutions was the economic stabilisation of crisis-ridden developing nations. Later these institutions embraced the ideology of international free trade, capital investment, and the rule of the market within the domestic sphere.

The 'Washington Consensus' later became a buzzword to describe and legitimate the broader enforcement of the strongly market-based approach to governing the economy emanating particularly from Britain and the United States, as described in the previous chapter.[1] How far the adoption of neoliberalism as the dominant ideology of the developed world influenced what happened in the Russian Imperium and its Eastern Bloc client states in the 1990s is debatable.

Certainly, however, the paradigm of what capitalism 'should be' in the 1990s, and what a capitalist economy 'necessarily' demanded, must have influenced what occurred. Those responsible for the rapid transformation of communism into capitalism were independent Western advisors, the International Monetary Fund, the United States government and, most importantly, President Yeltsin's Russian administration.[2]

Independent Western advisors certainly played an important part. But the specific method of conversion of Russian state communism to capitalism was decided upon by Russians. The dispersal of Russian wealth to private citizens was achieved through Russian central control. 'Thus, in less than ten years, from 1990

DOI: 10.4324/9781003382133-6

to 2000, post-communist Russia went from being a country that had reduced monetary inequality to one of the lowest levels ever observed to being one of the most inegalitarian countries in the world.'[3] Piketty comments that the fall of the Soviet Union led to an extreme form of hypercapitalism: 'a society of oligarchs engaged in grand larceny of public assets'.[4]

Crisis

There are inherent tensions within capitalist societies arising from the conflicting interests of the owners of wealth (capital) and workers whose subsistence and well-being are dependent on income from employment (labour). Polanyi, as we saw in Chapter 2, posed the tension as resulting from the interests involved in supporting and enlarging market relations and those interests seeking social and environmental protection from the ever-expanding market.

In the 1970s, the political philosopher Jürgen Habermas wrote of the 'legitimation crisis' of Western capitalism.[5] There was much truth in his argument. But democracies facing a legitimation crisis are able (to some degree via freedom of speech, freedom of association, and general elections) to change policy settings to adapt to new conditions and shore up, at least temporarily, a government's legitimacy. We saw that happen in both the social democratic and the neoliberal transformations. This is not the case with centralised autocratic regimes.

The transformation of the Soviet imperium from Stalinist communism to free market capitalism began and ended in Russia. But it was deeply rooted in prior structural crises throughout the Soviet empire stretching from Poland on the Baltic to Romania on the Black Sea. While the empire seemed to Western observers from the 1960s to the 1980s to be stable and invincible, its economy was ossifying. In Judt's account of the economic crisis there were three elements: an outdated model of economic development, inflexible and centralised economic management, and a permanent cycle of corruption.[6]

Since the end of the civil war in Russia in 1921, state-controlled production of primary industries had lifted Russia from an agrarian nation to an industrial power capable of defeating the Nazi Wehrmacht and threatening the Western democracies. But the Soviet system, which prevailed throughout Russia and the Eastern bloc, remained wedded to heavy industry. The system was based on an earlier model, 'redolent of Detroit or the Ruhr in the 1920s, or late nineteenth century Manchester'.[7] Unlike the democracies the system could not adapt to the new production and information technologies that were freeing companies for international competition. The Soviet bloc 'missed the switch from extensive to intensive high value production that transformed Western economies in the course of the Sixties and Seventies'.[8]

The Soviet system could neither produce good quality high-technology goods nor do so in sufficient quantity. For example, East Germany (the GDR: German Democratic Republic) was charged with manufacturing computers. The products were unreliable and outdated. By 1989, the GDR was producing one-fiftieth the number of computers manufactured in Austria, a country with half the GDR population. 'The GDR was spending millions of marks producing unwanted goods that were available at lower cost and in better quality on the world market'.[9]

The problem arose from the Soviet insistence on central economic planning. By the 1970s, the Soviet central planning agency had 40 departments for various branches of the economy, and 27 separate economic ministries. Central plans for every department and branch dictated precise numerical targets for everything produced and by when they had to be attained. Prices for products were centrally fixed so there was no connection between demand, supply, and price of goods. Administrators, secure in their jobs, were afraid of taking risks: 'factory foremen and managers took great pains to *hide* reserves of material and labour from the authorities'.[10] The result was simultaneous *waste* and *shortage* of what was wanted and needed.

The Polish journalist Ryszard Kapuscinski travelled through Russia shortly after the disintegration of the Soviet Union. He illustrates with fine observation of small details the terror perpetrated on the Russian people: for instance, the barbed wire industry churning out millions of miles of the stuff to imprison dissenters in the Gulags and prevent movement across the vast frontiers of the USSR. The shuffling of entire populations and ethnic groups across the Soviet territories. The steely gaze of a border guard trained to see spies in every face, spies who want to get in or get out.[11]

He writes of the atrocious environmental conditions in towns and cities. Of Kiev (now Kiiv) in Ukraine, he writes that one should not be misled by the enduring charm of the city:

> In many buildings, in entire housing complexes, people live very badly. The stairwells are filthy, the windowpanes broken, the outbuildings and courtyards dark because the lightbulbs have either been stolen or smashed. In many houses there is either no cold water or no hot water, or no water at all. Cockroaches and all other kinds of stubborn vermin are a universal plague.[12]

He writes of the destruction of the environments of Central Asia under orders from Moscow in order to industrialise the growth of cotton from which the cotton mafia got rich but millions of the cotton pickers went begging.[13] The waters of the two rivers, Syr Darya and Amu Darya, were plundered for cotton plantations, drying up the Aral Sea and killing the bordering town of Muynak: 'Today there is neither river nor sea. In the town the vegetation has withered; the dogs have died. Half the residents have left'.[14] One-third of the sea's surface area has become a desert poisoned by salt and artificial fertilisers.

A more scholarly and comprehensive account of environmental pollution in the Eastern bloc states (Albania, Bulgaria, Czechoslovakia, Hungary, Poland, Romania, and Yugoslavia), is provided in the edited collection by Carter and

Turnock.[15] The picture is more nuanced and complex than Kapuscinsky's reports from the Soviets. Pollution also crosses borders from Western countries. Nevertheless, pollution of land, air, and water was severe under Soviet rule.

Together with environmental pollution and destruction, the political outcome was permanent and entrenched corruption at every level throughout the system. Judt aptly observes that 'the absence of private property tends to generate more not less corruption. Power, position and privilege accrued from mutually-reinforcing relationships of patronage and clientelism' (producers at the workface serving the needs of their political controllers).[16]

The only parts of the economy that worked with reasonable efficiency by the 1980s were the high-technology defence industries and the 'second economy' (the black market in goods and services). At its peak, spending on the 'defence' industry occupied between 30 and 40 per cent of the Soviet economy.[17] The black market flourished. Judt cites the example of Hungary in the early 1980s where a mere 84,000 artisans operating exclusively in the private sector were estimated to be producing 'nearly 60 per cent of local demand for services, from plumbing to prostitution'.[18] By the 1960s, public support for the Communist regimes in Eastern Europe could no longer depend on the 'promise of socialism'. The public, unsatisfied with the socialist utopia 'tomorrow', wanted material abundance 'today'.[19]

Dissatisfaction with inefficient economies, environmental degradation, political nepotism, corruption, and severely restricted freedom of expression, civil organisation, and travel led to clandestine and sometimes open protest in the client states. Violent suppression by the Soviets in Hungary in 1956 and Czechoslovakia in 1968 simply fed and deepened the widespread resentment of Soviet communism.

Two disasters of political overreach and managerial inefficiency consolidated the sense that something had to change. The invasion and occupation of Afghanistan from 1979 resulted in devastating losses of soldiers and equipment and aroused an intransigent guerrilla war against the occupiers. The Red Army was humiliated in Afghanistan. Returned soldiers formed a festering bloc, turning to alcohol and far right ideology.

In April 1986, one of the four large graphite reactors of the nuclear power station at Chernobyl exploded, releasing over Europe more than a hundred times the radiation of the two nuclear bombs of Hiroshima and Nagasaki combined. The KGB knew about the shoddy equipment and operational deficiencies of the reactors, but the information was suppressed, and nothing was done. This was perhaps the most striking, but not necessarily the worst, of a series of environmental disasters perpetrated by the Soviets.

By this time, even though there was no organised dissident movement, voices challenging Soviet orthodoxy were beginning to be heard within Russia. The leadership of the Soviet Union had also changed. In 1985, Mikhail Gorbachev was elected Secretary General of the Communist Party of the Soviet Union. He was the first of the champions who drove the transformation from communism to capitalism. The question of ideology can best be discussed in relation to the Gorbachev period and that of his successor, the second champion, Boris Yeltsin.

Political Champions

By the 1980s, the rule of the old Communists at the head of the vast Party apparatus – 'an authoritarian gerontocratic bureaucracy'[20] – was coming to an end. Leonid Brezhnev, Yuri Andropov, and Konstantin Chernenko died in quick succession. A younger generation took over the leadership. In 1985, Mikhail Gorbachev took over as Secretary General of the Communist Party. Aged 54, he was younger by 20 years than his immediate predecessors.

Gorbachev had risen within the Party. He belonged to the optimistic generation following Khrushchev's denunciation of the sins (or 'cult of personality') of Stalin.[21] He believed that communism could be reformed in line with socialist principles. His goal was controlled modernisation, making production responsive to demand through the price mechanism (*perestroika*).

In order to prevent the party apparatus from obstructing his plans, he encouraged public discussion of selected topics (*glasnost*). He introduced contested elections for the Congress of People's Deputies and had himself elected President of the Supreme Soviet. He announced that Moscow would no longer use force to impose its will on the client states of Eastern Europe. In 1991, he gave the separate Republics of the Soviet Union the right of secession under a planned new federalist constitution of the Union.

Gorbachev was not a champion of capitalism, but his reforms opened the door for transformation to occur. Judt remarks, 'By introducing first one element of change and then another and then another, Gorbachev progressively eroded the very system through which he had risen'.[22] As Gorbachev's popularity waned, that of the Chairman of the *Russian* Supreme Soviet, Boris Yeltsin, ascended. Yeltsin, like Gorbachev, had risen through the Soviet apparatus, but he redefined himself as a specifically Russian leader. In 1990, from his power base in Moscow, he resigned from the Communist Party.

When the Supreme Soviet led by Gorbachev voted for a new decentralised Union with greater latitude for the member republics, the conservatives in the Party, including the Soviet Prime Minister, the Defence Minister, and the head of the KGB, prepared to replace Gorbachev and take over the leadership. While Gorbachev was on holiday at his Black Sea villa in 1991, the coup was launched led by Gennady Yanayev, Vice President of the Soviet Communist Party. Around 4,000 soldiers, 350 tanks, 300 armoured personnel carriers, and 420 trucks were sent to Moscow.[23]

Gorbachev refused to hand over power to an 'emergency committee'. Yeltsin denounced the plotters and led the resistance, calling on the crowds surrounding the Russian Parliament to defend democracy against the coup leaders and their armed forces. The coup failed and the failure emboldened Estonia and Latvia to declare independence. Days later, the remaining Soviet republics led by Ukraine followed suit. The Soviet Union, and with it, the Communist ideology simply disintegrated. There was a peaceful transformation without war or serious violence. But capitalism was still to come.

In a referendum in March 1991, Russians voted in favour of establishing a presidency and holding direct elections. In June of that year, the presidential election was held. Yeltsin was elected president of Russia with 56 per cent of the vote. This was the first, and some believe the only genuinely democratic election of a Russian president.

The means of creating a capitalist economy was termed 'shock therapy'. It was decided by Yeltsin and his economic advisors Yegor Gaidar and Anatoly Chubais, assisted by Western economists such as Jeffery Sachs, Andrei Shleifer, Per Anders Åslund, Richard Layard, and Stanley Fischer. The intention was to make it impossible ever again to return to communism. Private ownership and a free market had to be created *de novo* from a system of production almost entirely owned and controlled by the state – a truly daunting task. In the early 1990s, neoliberalism (or the Washington consensus) had become the West's dominant doctrine and that ideology was now applied in its crudest form.

Initially, the purchasers of Russia's wealth were offered a flat rate of income tax of 13 per cent with zero inheritance tax,[24] but the taxation regime was under constant debate in the parliament. According to one analysis, between 1993 and 1994, the ratio of taxes collected to GDP declined from 41 per cent to 36 per cent, although the percentage of GDP paid in taxes was already lower in Russia than in any of the Western market economies.[25] In the first quarter of 1996, only 56 per cent of planned tax revenue was actually collected.

Without waiting to establish the institutions in which markets are embedded in Western capitalist nations, the wealth of Russia embodied in all the organisations of production, large and small, was sold off in a voucher system followed by closed auctions rigged in favour of those who could assemble enough vouchers to make a bid.

The Ideology and Practice of 'shock therapy'

Piketty comments on the transformation of 'the land of the soviets' to the 'land of oligarchs and kleptocrats' – and we must now add the return to authoritarian rule greatly accelerated by the Russian war on Ukraine. In its totality and rapidity this was a transformation 'without precedent anywhere else in the world'.[26] While Piketty agrees that the transformation was in part the result of the failure of Soviet socialism for the reasons mentioned above, he argues that the Soviet failure cannot be an explanation for the particularly brutal capitalism that was subsequently imposed.

The choice of many forms of capitalism was available in 1990. The Scandinavian model, the West German model, even the post-war American or British models

– all were accommodations between market relations and social protection. None of these models was considered. There was no thought given to the actual history or the likely future of capitalism. Piketty observes that the extreme form of capitalism adopted demonstrates,

> the importance of crises in the history of inequality regimes. Depending on what ideas are available when a switch point arrives, a regime's direction may turn one way or another in response to the mobilizing capacities of the various groups and discourses in contention.[27]

We now need to consider in detail the nature of 'shock therapy'. In 1993, three economists who influenced the Russian government: Andrei Schleifer, Maxim Boycko, and Robert W. Vishny wrote a long article that subsequently became a book: *Privatizing Russia.*[28] Schleifer and Boycko are described as Russian-Americans. All the authors are well-respected and distinguished scholars at Harvard and the University of Chicago.

They write that in October 1991, privatisation was not at the top of the agenda of the Russian State Committee on the Management of State Property headed by Anatoly Chubais. A year and a half later, they write, 'privatization has become the most successful reform in Russia'.[29] In their article, the authors do not tell us how the crucial step was taken to bring privatisation to the top of the agenda.

Instead, in a discussion of the goals of privatisation, they invoke the liberal mantra which in the case of Russia was all too true: 'public enterprises are inefficient. They employ too many people, produce goods that consumers do not want, locate in economically inefficient places, do not upgrade their capital stock'.[30] But they continue: 'While these problems are particularly severe in eastern Europe, public enterprises throughout the world are conspicuous for their inefficiency, as well. This observation is no longer controversial'.[31] These contentions are 'no longer controversial' because they form a central plank of neoliberal ideology applied indiscriminately worldwide.

Public enterprises are inefficient, they argue, also because they tend to respond to political demands. Thus, depoliticisation is one of the goals of privatisation. Yet, as they add, 'privatization is always and everywhere a political phenomenon. The goal of governments that launch privatization always is to gain support for the reformist (or conservative) politicians'.[32]

It should be remembered that all permissible businesses were owned by the Russian state. According to the above authors, the privatisation programme involved three steps. First, firms were divided between those that could be sold for cash by local governments and those that would be placed in the mass privatisation programme. Most small-scale enterprises (shops and small businesses) were allocated to local governments to sell. These governments demanded their cut from the sales. Some were later sold for privatisation vouchers.

Secondly, larger firms were divided between those, subject to mandatory privatisation (light industries such as textiles, food processing, and furniture), those that required permission from the privatisation Ministry, those requiring government approval,[33] and those whose privatisation was prohibited.

Third, all large- and medium-sized firms were to be 'corporatized' – namely registered as joint stock companies with a board of directors, a corporate charter, and with all equity owned by the government. Divisions of such firms were allowed to split off and form separate enterprises.

Once corporatised, a firm's managers and workers could choose one of three options[34]:

1. Workers and managers to be given 20 per cent of the shares of the enterprise free. Top managers to be allowed to purchase 5 per cent of the shares at a nominal price. After privatisation workers and managers could get an additional 10 per cent of shares at a 30 per cent discount to book value (a kind of employee stock ownership plan).
2. Managers and workers to receive 51 per cent of the equity at 1.7 times the book value of the assets. Workers could pay in cash or with vouchers. The authors comment that, 'this, of course, represented a very low price relative to the market value of these assets in a highly inflationary environment'.[35]
3. Managers to be allowed to buy up to 40 per cent of the shares at very low prices if they promised not to go bankrupt. This option was imposed, but rarely used, at the behest of the managerial lobby in the parliament.

Having decided on one of the above options, the managers and workers could then submit a privatisation plan to determine how the remaining shares were to be sold.[36] The above authors say that the filing of privatisation plans was almost always voluntary. But the pressure to privatise and the monetary incentives to do so were plainly enormous.

The predominant method of privatisation was through the voucher system followed by auctions. In 1992, for a fee of 25 roubles one voucher for a share in the national wealth was issued to every Russian citizen including children. Each voucher had a denomination of 10,000 roubles, presumed to expire at the end of 1993. The vouchers were freely tradable and could be used as the sole permissible means of payment in auctions of shares of businesses to be privatised.

Because the vouchers were fully tradeable, investors with enough money were able to assemble blocks of vouchers giving the seller a premium for each sale. Blocs of vouchers sold on via chains of 'investors' trading with one another ended up as very large holdings in the hands of those who already had wealth or acquired it through voucher trading itself. The authors note: 'Apparently, investors willing to participate in voucher auctions do not face major problems in assembling large blocks of vouchers'.[37]

The above authors estimated that the total value of accepted vouchers came to about US$150 million.[38] They conclude that through trading in vouchers and sales at auction:

> That puts the total value of the Russian industry at about US$5 billion. It is possible to make these calculations differently and to come up with numbers as high as US$10 billion. The point, however, is inescapable: the entire Russian industrial complex is valued in voucher auctions at something like the value of one large 'Fortune 500' company.[39]

To further dramatise the situation, the market value of *ZIL,* the huge car and truck manufacturer employing 100,000 workers was found to be about US$16 million, 'including a large chunk of Moscow real estate'. 'The market values of *Uralmash* and *Permsky Motors*, two household names in Russian manufacturing, are $4 million and $6 million, respectively'. The low quality of Russian assets fails to explain their low market value by a factor of 100. 'Additional explanations are needed'.[40] A plausible explanation, in the authors' view, settles on expropriation of 'shareholders' by 'stake-holders'.[41]

The 'stakeholders' fall into three categories: employees of the state-owned businesses, the managers of those businesses, and foreigners. Employees acting collectively seek to increase their wages. Managers 'are likely to expropriate shareholder wealth through asset sales to their own privately held businesses and other forms of dilution. This theft by managers is probably the principal reason for the remarkably high capital flight from Russia'. Foreign investors evidently find the market prices in voucher auctions very attractive. However, 'At the same time, they do not usually openly challenge the managers, for fear of a political reaction. Indeed, they usually acquire their stakes through Russian intermediaries'.[42]

We need to look beyond the operation of a supposedly free market: to the design of 'shock therapy' itself. At first sight the distribution of vouchers equally to all Russians seems extremely egalitarian. But what did the possession of 10,000 roubles mean to a Russian household? Well, in 1993 the average Russian wage was about 23,000 roubles per month, so around 2 weeks' wages. That could be spent of course. But a more attractive option would be to sell your voucher to an 'investor' for a big mark up, perhaps for double the face value (in June 1993 still only worth US$2).[43]

It is fairly clear that the designers of the voucher system expected a process of accumulation of ownership to take place, in fact rather quickly, resulting in a limited number of owners of enormous blocs of vouchers. Now we have to consider the auction system. There was intense pressure to privatise the state firms quickly. There was nothing to stop bloc owners from collaborating prior to bidding at auctions and agreeing in advance who was going to bid for which asset, thus avoiding competition amongst them.

By 1995, Russian society was in turmoil. The Russian Mafia had become a powerful force. Citizens were being murdered on Moscow streets. The government was unable to pay its own officials' salaries or pensions because of the collapse of government revenue. The first round of privatisation was stalling, and the government was desperate for money. Yeltsin's approval rating in opinion polls fell to single digits. He faced an election in 1996 and an increasingly popular challenger who might restore communism.

So, Yeltsin initiated a further privatisation scheme that was known as 'loans for shares'. The scheme worked like this:

> The richest oligarchs loaned the government billions of dollars in exchange for massive shares of Russia's most valuable state enterprises. When the government defaulted on paying back the loans, as the schemers expected they would, the oligarchs would walk away with the keys to Russia's most profitable corporations. In exchange, the government would get the money it needed to pay its bills, privatization would keep moving forward – and, most importantly, the oligarchs would do everything in their power to ensure Yeltsin was re-elected.[44]

Between November and December auctions were held for 12 of Russia's most profitable enterprises including shipping, steel, and oil companies. The outcome was predetermined as to who would get what.[45]

There were those who favoured a more gradualist approach to the creation of a capitalist market in Russia, including Jeffrey Sachs, who argued that structural change in the economy is a task for one or two decades. Douglas North advised that while rules may be changed overnight, informal norms change only gradually, and simply transferring the formal political and economic rules to Eastern European economies is not a sufficient condition for good economic performance.[46]

Joseph Stiglitz argued that, without first building a sound institutional framework, economic policies would falter. Corruption was a major problem. Between 1992 and 1999 capital flight from Russia amounted to at least US\$ 45–50 billion each year.[47] As much as 30 per cent of that occurred illegally. Critics accused the Western investors and oligarchs of money laundering.[48] Stiglitz concluded that:

> The radical reformers in Russia were trying simultaneously for a revolution in the economic regime and in the structure of society. The saddest commentary is that, in the end, they failed in both: a market economy in which many old party apparatchiks had simply been vested with enhanced powers to run and profit from the enterprises they formerly managed, in which former KGB officials still held the levers of power. There was one new dimension: a few new oligarchs, able and willing to exert immense political and economic power.[49]

Stiglitz suggests that the transformation was entirely in keeping with Soviet culture and revolutionary change in general: the French Revolution of 1789, the Paris

Commune of 1871, the Bolshevik Revolution of 1917 and China's cultural revolution. 'Each produced its own Robespierre, its own political leaders who were either corrupted by the revolution or took it to extremes'.[50] With a kind of historical inevitability, the capitalist revolution threw up Vladimir Putin.

A former KGB officer from St Petersburg, Putin was nurtured by Boris Yeltsin, becoming Deputy Chief of Yeltsin's presidential staff. In 1998, he was appointed director of the FSB – the federal security service (the KGB's successor). Yeltsin wanted Putin to be his successor in the presidency. He became acting president in 1999 after Yeltsin resigned. His first act was to exonerate Yeltsin from all charges of corruption. Elections were held in March 2000. Putin was elected President with 53 per cent of the vote.

Commenting on Russia's turn to democracy following the dissolution of the Soviet Union, Aleksandr Solzhenitsyn said in an interview in 1992,

> The system that governs us is a combination of the old nomenklatura, the sharks of finance, false democrats, and the KGB. I cannot call this democracy – it is a repugnant, historically unprecedented hybrid, and we do not know in which direction it will develop ... but if this alliance will prevail, they will be exploiting us not for seventy, but for one hundred and seventy years.[51]

Consequences

Shock therapy is one of those memes designed to legitimise the unacceptable: like, 'you can't make an omelette without breaking eggs', 'tough love', 'spare the rod and spoil the child'.

The immediate consequences of the transformation were disastrous for Russia. Industrial output fell and unemployment grew. As the economy collapsed hyperinflation was released. In the following 4 years consumer prices rose almost two thousand times from 1990. Many Russians were left destitute.

The new class of industrialists and bankers had come into immense power through the plunder of Russia's industrial wealth and corrupt deals with the government. Unable to achieve consensus for the reform programme, Yeltsin turned to rule by presidential decree, cementing deals with the powerful newly enriched oligarchs.

Inequality and poverty in Russia surged. A Russian study (from the Higher School of Economics) has shown that Russians living in poverty grew from 2 per cent in 1987 to 50 per cent in 1993–1995.[52] Piketty records that, in 1980, the top centile's share of income in the Soviet Union averaged 5 per cent (of total national income), compared with between 5 per cent and 10 per cent in Europe. After 1990, the top centile's share of income in Russia rose to between 20 per cent and 25 per cent. The top centile's (top 1 per cent) share of total private wealth also grew from about 15 per cent in 1980 to around 40 per cent in 2000, higher than that of the USA, the UK, and China.[53]

With the loss of state ownership of almost all the productive enterprises, the Russian government was rendered economically powerless. The immense wealth of the oligarchs in an infant democracy placed political power in their hands, power to buy electoral support, power to sway the electorate, power to protect and further enrich themselves. But the state still possessed the power of violence. Putin's government began to wage war against corrupt individuals, criminal syndicates, and the oligarchs. Between 1997 and 1999, it has been reported, at least a dozen 'legislative aids' were murdered, most of whom were found to have criminal ties.[54]

In Vladimir Putin's first period as President, the economy was stabilised and grew by an average of 7 per cent per year (with a fivefold increase in the price of gas and oil).[55] But he allegedly began to use murder and imprisonment to force his enemies to back him. Although technically under the rule of law, the justice system was increasingly controlled by the political arm of the state.

William Browder chronicles in detail how his Russian lawyer Sergei Magnitsky was imprisoned and beaten to death in 2009 after he agreed to testify against Russian ministry officials who had conspired to steal US$230 million of Russian taxes paid by a major hedge fund.[56]

In 1997, an economist and St. Petersburg deputy mayor, Mikhail Manevich was shot and killed on a busy street. In November 1998, Galina Starovoitova, a Soviet dissident and one of Russia's most popular politicians, was gunned down outside her St. Petersburg apartment. Starovoitova, a supporter of Andrei Sakarov (a well-known dissident from Soviet days), was planning to run for president in 2000. In 1999, Victor Novosysolov, another St. Petersburg legislator, was decapitated by a bomb in his car. In 2006, Journalist Anna Politkovskaya, a leading critic of the Chechen war and opponent of Putin, was shot dead outside her Moscow apartment. Famously, Boris Nemtsov, a towering figure in Russian post-Soviet politics and a biting critic of President Vladimir Putin was gunned down on the steps of the Kremlin.

Poisoning was a less successful weapon of assassination. In 2006, Alexander Litvinenko, a secret-police whistle-blower, was killed by polonium in London. In 2008, Karinna Moskalenko, fell ill in Strasbourg with mercury poisoning but survived. But Yuri Shchekochikhin, her colleague at the investigative weekly Novaya Gazeta, died in a Moscow hospital.[57] Moskalenko was one of the lead lawyers in the defence of the oligarch Mikhail Khodorkovsky.

In March 2018, Sergei Skripal, a former Russian military officer and double agent for the British intelligence agencies, and his daughter, Yulia, were poisoned in the city of Salisbury, England, with a Soviet nerve agent, Novichok. They survived, but a man and a woman died after picking up the perfume bottle containing the poison. In 2020, Alexei Navalny was the victim of attempted poisoning with Novichok rubbed into his underpants. He survived and returned to Russia where he was immediately arrested, charged, and imprisoned. 'Anti-corruption campaigner Alexei Navalny has long been the most prominent face of Russian opposition to President Vladimir Putin'.[58]

In the turmoil caused by the capitalist transformation, Russia led by Putin and his St. Petersburg clique, turned to 'the fascist solution', crushing all opposition. Of the 1930s, Polanyi writes with alarming resonance today that, 'The fascist solution of the impasse reached by liberal capitalism can be described as a reform of market economy achieved at the price of the extirpation of all democratic institutions'. It offered an escape from institutional deadlock – 'yet if the remedy were tried, it would everywhere produce sickness unto death. That is the manner in which civilizations perish'.[59]

The term 'fascism', overused and misused during the twentieth century, originated with Mussolini's Italy, but Polanyi argues that the same mix of liberal capitalism with the extirpation of democracy could be found in numerous regimes and cultures responding to the crisis of liberalism. In fact, 'there was no country immune to fascism, once the conditions for its emergence were given'.

A country approaching the fascist phase, Polanyi writes, showed symptoms such as 'the spread of irrationalistic philosophies, racialist aesthetics, anticapitalist demagogy, heterodox currency views, criticism of the party system, widespread disparagement of the "regime" or whatever was the name given to the existing democratic setup'.[60] Today these symptoms are appearing again in many nations, not least in the United States – resulting in the January 2021 armed assault on the Congress instigated by Donald Trump.

The final seal was placed on fascist Russia by the invasion of Ukraine, an infant democracy also struggling with oligarchs, and a fraternal Slavic state posing no threat to Russia. Some argue that the Ukraine War was necessary to maintain Putin's hold on power just as his popularity was slipping.[61] As with Afghanistan, the gerontocratic regime badly miscalculated its power to enforce its will on a resistant people. Russia has been humiliated.

The long-term consequences are still playing out. From the coup against Gorbachev onwards two things became clear: an intense resentment within the political class about the loss of the Soviet imperium, and a focus henceforth on the identity and ideological importance of Russian nationalism. Under Putin democracy has been increasingly dismantled. The war against Ukraine has demonstrated that Russia has become a fascist power, weakened economically and militarily, shunned by Western capitalism, but nuclear-armed. The Cold War with the West has recommenced. Europe, if not quite united, has drawn itself together in opposing Russia.

Yet it is too easy to single out Russia and its regime as a scapegoat for what has deeper systemic causes. A distinguished Russian political scientist working in the USA, Igor Okunev, writes that, after the collapse of the Soviet Union, Russia made attempts to keep pace with the developed world. Russians viewed their position as a victory over the authoritarianism of communism. The West, on the other hand, interpreted the breakup of the Soviet Union 'not as Russians' victory over the Communist regime, but as their defeat in the Cold War. Moscow's attempts to remind the West about equitable partnership were basically dismissed with a polite smile of the winner'.[62]

Okunev continues:

Russia's victory in World War II became the central idea of his [Putin's] presidency. Russia's self-sacrifice for the sake of peace – liberation of the world from Nazism – is now the country's key fetish which, as new ideologists believe, is uniting the old and contemporary Russia. It is the lost national idea that is consolidating people and marking new boundaries between 'us', as a new post-Soviet community, and 'them', as our enemies.

The Manichaean memes of 'democracy versus autocracy': the Russian nation and its enemies, are simplistic and diversionary from systemic failure. As Piketty observes, it is legitimate to ask why Western nations were 'so uninterested in the origins of Russian wealth and so tolerant of the mass misappropriations of capital'.[63] To continue: Russia's post-Communist trajectory reflects in part the failure of social democracy and participatory socialism to develop new ideas and a workable plan for international cooperation in the late 1980s and early 1990s. The misreading of the situation of Russia in the world, together with blindness to the failure of liberal democracy, has had disastrous consequences for peace in the international order.

Notes

1 The term was created by John Williamson, an economist with the Institute of International Economics. Williamson proposed a set of ten policy principles: 1. Avoidance of large fiscal deficits relative to GDP. 2. Redirection of public spending from subsidies to pro-growth services that would alleviate poverty such as primary education, health care, and infrastructure. 3. Tax reform to broaden the tax base. 4. Interest rates determined by the market. 5. Competitive exchange rates. 6. Liberalisation of international trade. 7. Liberalisation of direct foreign investment. 8. Privatisation of state enterprises. 9. Abolition of regulations that restrict trade and markets except for those relating to safety, environmental and consumer protection, and prudential oversight of financial institutions. 10. Legal security for property rights. https://en.wikipedia.org/wiki/Washington_Consensus#Original_sense:_Williamson's_Ten_Points. (accessed 20/10/2022).
2 Aslund A. (2007a).
3 Piketty, T. (2020) p. 597.
4 Ibid. p. 595.
5 Habermas, J. (1975).
6 Judt, T. (2005) pp. 576–581.
7 Ibid. p. 578.
8 Ibid. p. 578.
9 Ibid. p. 578
10 Ibid. p. 579.
11 Kapuscinski, R. (1994).

12 Ibid. p. 275.
13 Ibid. p.259
14 Ibid. p. 261.
15 Carter, F.W. and Turnock, D. eds (1993).
16 Judt does not cite corruption in free market capitalism. Piketty provides ample empirical evidence of power, position, and privilege accruing from wealth (Piketty, T., 2022).
17 Judt, T. (2005) p. 592.
18 Ibid. p. 579.
19 The beginning of scepticism about 'tomorrow's' utopia appears in the epic war novel
   Stalingrad by Vasily Grossman.
20 Judt, T. (2005) p. 594.
21 *The Optimists*: A Russian TV Series that captures the spirit of some young apparat-
   chiks in the Khrushchev era looking to a more humane and open future overshadowed
   by the insidious power of the KGB (https://www.imdb.com/title/tt5905344/).
22 Judt, T. (2005) p. 603.
23 (https://en.wikipedia.org/wiki/1991_Soviet_coup_d%27%C3%A9tat_attempt#19
   _August)
24 Piketty, T. (2020) p. 603.
25 https://factsanddetails.com/russia/Government_Military_Crime/sub9_5a/entry-5196
   .html#chapter-6 (accessed 07/11/2022).
26 Piketty, T. (2020) p. 597.
27 Ibid. p. 605.
28 Shleifer is a 'Russian-American' Professor of Economics at Harvard University,
   Boycko, likewise Russian-American, at the Economics Department of Harvard and
   a senior fellow of the Watson Institute for International and Public Affairs at Brown
   University. Vishny is the Distinguished Professor of Finance at Chicago University.
   Privatizing Russia was first published as a long article in 1993 by BPEA (Baring
   Private Equity Asia) at the time an affiliate of Barings Bank. Boycko was Deputy Prime
   Minister and Deputy Chief of Staff to President Yeltsin. 'Mr. Boycko played a principal
   role in the design and implementation of the Russian voucher privatisation program
   in 1992–1994 and macroeconomic stabilization program in 1995' (https://daviscenter
   .fas.harvard.edu/about/people/maxim-boycko, accessed 01/11/2022). Schleifer was 'a
   direct advisor to Anatoly Chubais, then vice-premier of Russia' (https://en.wikipedia
   .org/wiki/Andrei_Shleifer. Accessed 02/11/2022).
29 Boycko M, Schleifer, A. and Vishny, R.W. (1993) p. 139.
30 Ibid. p. 141.
31 Ibid. p. 142.
32 Ibid. p. 147.
33 Ibid. p. 149 'Major firms in most strategic industries, such as natural resources and
   defence, could only be privatized with the agreement of the entire government'.
34 Ibid. The following is from pp. 149–150.
35 Ibid. p. 150.
36 Ibid. p. 150.
37 Ibid. p. 155.
38 Since the rouble was in 1993 freely convertible on foreign exchanges, the calculation
   is based on the US dollar value placed by the Russian government on the total of all
   state-owned enterprises.
39 Ibid. p. 159.
40 Ibid. p. 160.
41 Noting the US$15 billion of capital flight from Russia in 1992.
42 Ibid. p. 171.
43 According to the New York Times of June 1993: https://www.nytimes.com/1993/06
   /01/business/ruble-reaches-record-low-dropping-to-1024-to-dollar.html   (accessed
   02/11/2022)

44 Rosalsky, G. (2022).
45 Ibid.
46 Aslund, A. (2007b) p. 50.
47 Sakwa, R. (2008).
48 Bedirhanoğlu P. (2004) pp. 19–41.
49 Stiglitz J. (2002) p. 163.
50 Ibid.
51 Cited by Kapuscinsky K. (1994) (see also: https://positionpapers.ie/2022/06/editorial -june-july-2022/). 70 years was roughly the duration of the Soviet Union.
52 https://iq.hse.ru/en/news/527970668.html#. (accessed 03/06/2022)
53 Piketty (2020, p. 584) comments that the Soviet regime organised inequalities in other ways through in-kind benefits and privileged access to certain goods. He notes that the regime was characterised by mass incarceration of whole classes of people as well as the other effects of totalitarian control. (Inequality of wealth is recorded in p. 671 Figures 13.8 and 13.9).
54 https://factsanddetails.com/russia/Government_Military_Crime/sub9_5a/entry-5192 .html (accessed 04/06/2023)
55 Sakwa, R. (2007) Chapter 9.
56 Browder, B. (2016). Browder is an American-British investment banker and CEO of Hermitage Capital Management, an advisor to the Hermitage Fund, specialising in Russian markets, and benefiting thereby. The Fund is headquartered in Guernsey with offices in the Cayman Islands. Hermitage Capital Management is reported as helping to expose high-profile cases of corruption in Russia's largest company Gazprom between 1998 and 2000.
57 'The symptoms of his illness fit a pattern of poisoning by radioactive materials and were similar to the symptoms of Nikolai Khokhlov (1957), Roman Tsepov (2004), and Alexander Litvinenko (2006)' https://en.wikipedia.org/wiki/Yuri_Shchekochikhin. (Accessed 10/11/2022)
58 BBC World News 07/10/2022, https://www.bbc.com/news/world-europe-16057045 (accessed 08/11/2022).
59 Polanyi, K. (1944) p. 245.
60 Ibid. p. 246.
61 Knott, M. (2022) 'Putin "needs war" to stay in power'. *The Age* Melbourne, 08/11/2022. Citing Sir Lawrence Friedman, emeritus professor of war studies, Kings College London.
62 Okunev, I. (2014).
63 Piketty, T. (2020) p. 605.

References

Aslund, A. (2007a) *Russia's Capitalist Revolution: Why Market Reform Succeeded, and Democracy Failed.* Washington, DC: Peterson Institute for International Economics.

Aslund, A. (2007b) *How Capitalism Was Built: The Transformation of Central and Eastern Europe, Russia, and Central Asia.* Cambridge: Cambridge University Press.

Bedirhanoğlu, P. (2004) 'The Nomenklatura's passive revolution in Russia in the neoliberal era', in McCann, L. ed. *Russian Transformations: Challenging the Global Narrative*, London and New York: Routledge.

Boycko, M., Schleifer, A., and Vishny, R.W. (1993) *Privatizing Russia*, Brookings Papers on Economic Activity, 2. https://scholar.harvard.edu/files/shleifer/files/privatizing _russia.pdf (accessed 02/11/2022).

Browder, B. (2016) *Red Notice, How I Became Putin's Number 1 Enemy*, New York: Simon and Schuster.

Carter, F.W. and Turnock, D. eds. (1993) *Environmental Problems in Eastern Europe*, London and New York: Routledge.

Habermas, J. (1975) *Legitimation Crisis*, tr. T. McCarthy, Boston: Beacon Press.

Judt, T. (2005) *Post War, A History of Europe Since 1945*, London: Penguin Books.

Kapuscinski, R. (1994) *Imperium*, tr. K. Glowczewska, New York: Alfred Knopf.

Knott, M. (2022) 'Putin "needs war" to stay in power'. *The Age* 08/11/2022.

Okunev, I. (2014) 'Different realities, The Crimean Crisis exposing the decline of the world order', *Russia in Global Affairs*, 12(2), 74–81.

Piketty, T. (2020) *Capital and Ideology*, Cambridge: Belknap Press of Harvard University.

Piketty, T. (2022) *A Brief History of Equality*, Cambridge: Belknap Press of Harvard University.

Polanyi, K. (1944, 2001 edition) *The Great Transformation, The Political and Economic Origins of Our Time*, Boston: Beacon Press.

Rosalsky, G. (2022) 'How shock therapy created Russian oligarchs and paved the path for Putin', Washington, DC: National Public Radio. https://www.npr.org/sections/money /2022/03/22/1087654279/how-shock-therapy-created-russian-oligarchs-and-paved-the -path-for-putin, (accessed 20/10/2022).

Sakwa, R. (2007) *Putin: Russia's Choice*, London and New York: Routledge.

Sakwa, R. (2008) *Russian Politics and Society*, London and New York: Routledge.

Stiglitz, J. (2002) *Globalization and Its Discontents*, London and Allen Lane, Penguin Books.

6

THE STATE AND CIVIL SOCIETY IN TRANSFORMATIONS

Introduction

In the preceding chapters, we have emphasised the essential part played by political actors and state institutions in creating social transformations. But activists in civil society, generating mass public movements, have also played a leading role in transformations. In recent years, we have seen mass activism formed for gender equality (the #MeToo movement), against racism (the Black Lives Matter movement), and action to stop global heating (School Strike for Climate and Extinction Rebellion).

How important have mass movements been historically in generating social transformation? How has the invention of democracy with universal adult suffrage affected the significance of mass movements? Karl Polanyi wrote about the 'double movement' in society: to create a market-based society, and to protect society from its devastating social effects. In characterising both sides of the equation of modern capitalism as 'movements' Polanyi is insisting that both were socially created within 'civil society'.

The fundamental struggle he identified was between those supporting and amplifying a market society, and those demanding protection from its excesses with an appeal for fairness, justice and democracy. Through the twentieth and into the twenty-first centuries that analysis of social movements remains true.

There is something like a continuum between day-to-day politics with its daily conflicts and controversies, and more deep-seated conflicts of interest and societal shifts. The social movement supporting 'marketisation' was a shadowy alliance of economic elites in government circles, well-funded private 'think tanks' (with their 'market' ideologies), and the owners of wealth who now include a substantial section of the middle class. Social movements for democracy and social justice are more diverse.

DOI: 10.4324/9781003382133-7

The Struggles for Democracy in Western Europe

The mass movements for 'social protection' of the nineteenth and twentieth centuries focused on the establishment, or enlargement and enrichment, of democracy. But, as Polanyi records, contra-Marx, the principle of social protection was also pursued by actors who did not belong to the working class: the Justices of Speenhamland, industrialists like Robert Owen, Liberals such as Franklin Roosevelt in the USA and, in the UK, David Lloyd George, William Beveridge, Maynard Keynes, and leaders of the Labour Party such as Clement Attlee.

It's notable that none of the above is a woman. Polanyi did not address one of the most important and ultimately successful movements for democracy from the nineteenth to the twentieth centuries: the women's movement. Why? Perhaps because the women's movement was not directly relevant to social protection from the unfettered market. Or perhaps because he was a man. As feminists have long argued, men who have written the history of society and politics and the environment have been quite unaware of an enormous lacuna in their thought: women, their subordination and subjugation, and their struggle to be accepted as political actors.

Women writers in the mid-nineteenth century drew attention to the condition of society. Elizabeth Gaskell (for one example) in her novel *North and South* paints an incisive portrait of both the dynamism of industry in the North of England and the crushing paternalism of the rural South (the Hampshire of Jane Austin).

In the novel, the Northern leaders of the cotton industry, devoted to the liberal ideology, ruthlessly suppress attempts by workers to organise collectively to improve their working conditions – always in fear of workplace illnesses and starvation. Yet those workers are also depicted as sharing a working life, however dangerous to their health and with miserable wages, that offers them collective self-respect and hope.

At a poignant moment in the novel the working-class hero, Nicholas Higgins, sacked by his boss for trouble-making, is thinking about moving to the rural South to find work. The middle-class woman at the centre of the romantic plot, Margaret Hale, urges him not to go. She tells Higgins of the mindless, unchanging drudgery of the farm labourer: a working life in a paternalistic society with no future, a life of 'living on the rates' (the Speenhamland system discussed by Polanyi).

As Polanyi argues (Chapter 2), the struggle of working people to improve the conditions of their lives and avoid starvation and misery was a crucial civil societal movement in the nineteenth century focused on work and industry. The organisation of factory production helped nurture the trade union movement. The mass movements in the nineteenth and early twentieth centuries, involving the trade union movement itself, were aimed at a political transformation: for democracy, for working people to gain control of the state.

Gradually the women's movement, through continuous organised struggle, won the right to vote in parliamentary elections: New Zealand, the first, in 1893 (though they could not stand as candidates until 1919), the colony of South Australia in 1894, the federal Commonwealth of Australia in 1902, Finland in 1906, the US state of Washington in 1910, California in 1911 (followed by other states in successive years), Denmark in 1915, Uruguay, Russia briefly, Ukraine, Sweden, and Britain in 1917, France, Italy, and Japan in 1945. Switzerland only in 1971 (and in one Canton 1991).[1]

The struggle for democracy spread over the whole world during the twentieth century. The defeat of the Fascist Axis of Germany, Italy, and Japan in the Second World War returned democracy to half of Europe. The end of colonialism in Africa brought democracy to many emerging states including eventually South Africa. Fascist dictatorships in Spain, Portugal, and Greece were ousted. Authoritarian regimes in South America – Argentina, Brazil, Chile – were replaced by democratic institutions. In Asia, Indonesia, the Philippines, South Korea, and Taiwan turned to democracy.

Political scientists Bastian Herre and Max Roser have covered the progress and the variety of democracies that have developed in the twentieth century.[2] The picture, though complex, is inescapable: movements for democracy have had extraordinary momentum and success. The twentieth century movement for democracy, as Judt insists, was fed by the memory of the costly but ultimately successful struggle between democracy and fascism, the moral legitimacy of the welfare state and the expectation of social progress which a strong and effective national government could deliver. 'It was the Master Narrative of the twentieth century'.[3]

Civil Movements against Repressive Regimes

The examples of struggles in Eastern Europe from within civil society described by Tony Judt in *Post War* demonstrate the importance of civil opposition even within the most repressive regimes.

The totalitarian dictatorships were still, in the 1980s, experienced by everyone every day. This experience generated a festering resentment in the Soviet's client states long before the collapse of communism in Russia. That resentment bred a generation of intellectual 'dissidents' (not as self-described but the term used in Western commentaries). The push back against suppression varied among different countries.

The common theme amongst dissident intellectuals in Czechoslovakia, Hungary, and Poland was the reconstitution of civil society embedding human rights and escape from under 'the unbearable tyranny of the citizen' that the state was supposed to embody.[4] After the Warsaw Bloc invasion of Prague in 1968 it became clear that communism could not be reformed. So, the dissidents began to speak beyond the current governments to 'suggest to the nation – by example – how it might live'.[5]

In Poland, following a period of strikes and demonstrations in 1976, a Committee for the Defence of Workers (KOR) bridged the divide between intellectuals and workers to put together 3 years later a 'Charter of Workers Rights' to *assert* in actuality, rather than simply *postulate* theoretically, an autonomous civil sphere. The Czech dissident Vaclav Havel urged his followers not to argue with those in power. The only thing that made sense in the circumstances of the time was to live in truth, 'to live *as if* one were truly free'.[6]

In January 1977, a group of Czech citizens signed a document, *Charter 77*, criticising their government for failing to implement the human rights provisions of the Czech Constitution, the Final Act of the 1975 Helsinki Accords, and the United Nations covenants on political, civil, economic, and cultural rights.[7] The aim of these acts of defiance was to overcome the cynical indifference of the public to public affairs among their fellow citizens.

In some Eastern bloc countries, for instance Poland and East Germany, dissent found a place in the churches. After Russia, East Germany – the German Democratic Republic – was perhaps the most authoritarian regime in Europe. Yet 'World Peace', proclaimed as an aim of communism, became a way of connecting with anti-capitalist 'Peace' movements in the West. The politically safe aim of 'Peace' was adopted by the Protestant *Bund der Evangelischen Kirchen* as a way in which the question of rights and liberties could, at least indirectly, be raised. For how can peace be hoped for in a country where, as the Czech dissident Vaclav Havel put it, the state is at permanent war with its own citizens?

The East German regime, from 1964, allowed an exemption from the 18 months of military service on conscientious grounds for those who joined an alternative labour unit called the *Bausoldaten*. Although taking this exemption could prove an economic handicap in later life, thousands of men took this option which created an informal network of peace activists. Thus, dissenting East Germans found a way, through the Peace Movement, of communicating with dissident opposition elsewhere in the Communist bloc.

The work of the few courageous 'dissident' individuals and networks prompted violent repression from the Communist regimes which in turn fed disillusionment and in time wider support for human rights. The emerging environmental catastrophe in Eastern Europe (mentioned in the preceding chapter) raised increasing fear and anger directed at governments, for 'under socialism it was the state that polluted'.[8]

Censorship was general and extensive. Thousands of men and women were denied a public voice. Many ideas, themes, and concerns could not be mentioned. But self-censorship was worse. Intellectuals and artists seeking public support were always tempted to trim and hedge their work in anticipation of likely official objections. In the face of constant repression, the default position for the vast majority of Eastern Bloc people trying to make a life for themselves and their families was studied apathy and cynicism.

The *intellectual* opposition in the Eastern Bloc did not by itself change much. The intellectuals talked mostly to other intellectuals rather than the population at large. Nevertheless, what Judt calls 'the power of the powerless', (borrowing the title of his chapter from an essay by Havel) flowed beneath the surface of societies in various ways and with different means.[9] Or as, Judt puts it:

> By forging a conversation about rights, by focusing attention on the rather woolly concept of 'civil society', by insistently talking about the silences of Central Europe's present and its past – by moralizing shamelessly in public, as it were – Havel and others were building a sort of 'virtual' public space to replace the one destroyed by Communism.[10]

Democracy, which eventually became the inevitable demand was, as ever, a means to an end, or rather many ends: freedom of the individual from arbitrary state power, freedom of expression and movement, freedom of association, and also social justice, fairness of outcomes, abolition of unacceptable inequality and poverty, abatement of environmental degradation, corruption, and limitation of the concentration of power.

No authoritarian regime that relies for its existence on 'making war' on the society it rules can survive indefinitely. Awareness of injustice seeps in. Leaders become irrelevant and are replaced. Economies are destroyed. Eventually power at the top speaks: 'This cannot continue'. However, even after Gorbachev had detached the Soviet Union from direct oversight of the client states – removing military intervention as the ultimate threat, the Eastern Bloc authoritarian regimes remained with 'a massive repressive apparatus'.[11]

In Poland, the independent trade union *Solidarity* gained strength, leading labour protests and strikes. Despite being banned at home, the union's organisation abroad was allowed in 1985 to negotiate with the International Monetary Fund to admit Poland. Judt reports that 'The "counter-society" theorized by the Polish historian Adam Michnick and others a decade earlier was emerging as a *de facto* source of authority'.[12] Plans were drawn up outside the Party apparatus for an autonomous private business sector. In 1988, a massive movement of industrial action with stoppages and occupations of industrial plant forced the Communist Party into irrelevance. A year later on February 6th the Party recognised Solidarity as a negotiating partner, a move which led 3 months later to the legalisation of independent trade unions.

The movement to democracy followed. A general election was held on 4 June 1989 with half of the seats in the Lower House of Parliament (the *Sejm*) reserved for Communists and a free vote for the whole of the Upper House (the *Senat*). In a shocking result *Solidarity* won 99 out of the 100 Senate seats, and all of the available seats in the *Sejm*.[13] The Communist Party was destroyed. Without backing from the Soviet Union, the option of declaring the election void and invoking martial law was off the table – under specific instructions from Gorbachev.

Communism in Poland was finished. It was, Judt notes, a long goodbye. Most of the actors had been on the political stage for many years.

> For all the strength of the Catholic Church, the countrywide popularity of *Solidarity*, and the Polish nation's abiding loathing of its communist rulers, the latter clung to power for so long that their final fall came as something of a surprise.[14]

In Hungary, the retreat from Communism was an 'in house' affair. Two decades of 'ambiguous tolerance' had obscured the limits of external dissent. The memory of the violent suppression of the revolution in 1956 also advised caution. Independent 'civil society' organisations that appeared in the 1980s were mostly concerned with environmental issues and the mistreatment of Hungarian minorities in Romania – safely within the sphere of Communist tolerance. Dissent was confined within the Party.

Real political change, however, began after 1988. A younger generation of Communists, enthusiastic for Gorbachev's Soviet reforms, succeeded in deposing the ageing Janos Kadar who had ruled Hungary since the 1956 revolution. Kadar represented the 'official lie' that reform that had occurred up to 1956 was nothing more than bourgeois counter-revolution. He also embodied the conspiracy of silence concerning the 1958 abduction of the Party leader, Imry Nagy, and his secret trial and execution. The removal of Kadar therefore had profound political significance.

In 1989, the Communist Parliament under the new leadership passed a series of measures recognising the right of free assembly and permitting a multi-party system, including the new dissident party *Fidesz* (or 'Young Democrats') led by a young firebrand, Victor Orbán. Of great symbolic importance was the act of recognition by the government of the events surrounding the death of Imry Nagy. His remains and those of four of his colleagues were exhumed and ceremoniously reburied. The 1956 Revolution was properly recognised. Its martyrs were proclaimed heroes. Some 300,000 Hungarians lined the streets to watch the event. Millions more watched on television.

The second Hungarian Revolution of 1989, writes Judt, had two features. It was the only transformation of communism to become a genuine multi-party democracy carried out entirely from within. And, whereas in Poland and Czechoslovakia the transformations were 'largely self-referential' (that is, without impact on other states), the Hungarian transformation played a vital role in the unravelling of communism in East Germany.[15]

In the 1980s, the West German government's stance of *Ostpolitik* had sought to reduce tensions with East Germany and facilitate human and economic communications across the border. The assumption was that the East German regime would continue indefinitely: a comfortable and self-serving illusion according to Judt.[16]

With the arrival of Gorbachev in Russia and the withdrawal of military intervention by the Warsaw Pact in support of regimes in the client states, dissidents became more confident, and public expressions of opposition against the regime began. In 1987, demonstrators against the Berlin Wall, praising Gorbachev, were violently dispersed. The following year, a demonstration commemorating the 1919 murders of Rosa Luxemburg and Karl Liebknecht was suppressed and a hundred demonstrators were imprisoned or expelled. The East German Party leader, Erich Honecker, was unbending. The blatant rigging of municipal elections in May of 1989, which returned a vote of 98.85 per cent for government candidates, elicited outrage from priests, environmental groups, and even critics within the government party.

In May of that year, Hungary relaxed the control of public expression and movement within the country. Although the border with East Germany remained formally closed, the Hungarian government removed the electrified border fence. East Germans started to believe that they had a choice. By July 1989, some 25,000 East Germans had chosen to take holidays in Hungary. The flood of citizens leaving East Germany grew. Hungary then opened its borders. Within hours 22,000 East Germans had crossed through.

Immediately, East German dissenters in East Berlin formed new movements, 'New Forum' and 'Democracy Now'. A protesting crowd of 10,000 in the East German city of Leipzig denounced the regime's refusal to countenance reform. Gorbachev himself came to visit and publicly advised the government not to delay reform. In Leipzig and other cities regular demonstrations and vigils for change were held, gathering crowds of thousands.

On October 18th an internal coup within the Party led by Egon Krenz, Honecker's deputy, removed the long-serving leader and set about a cautious *perestroika* reform with more liberal laws governing travel. On November 9th at a televised news conference an East German member of the Politburo Günter Schabowski announced that East German citizens could now travel abroad without having to apply for permits. At that moment the Berlin Wall was officially breached. The East–West border was opened. Over Christmas of 1989 some 2.5 million East Germans visited the West. The East German regime was, astonishingly, taken by surprise – still believing its own propaganda. Krenz, perhaps anticipating what was coming to him, said that the fall of the Berlin Wall was the worst night of his life. In 2000, he was sentenced to six and a half years in prison for manslaughter for crimes committed under communism.

These examples demonstrate the importance and variety of movements in civil society in the process of transformation. Social struggles all had one theme in common: to bring those in power in the state under the control of the public. But what of civil society movements in established democracies?

Social Movements in Democracies

In the democracies social struggles have taken three forms at overlapping moments. Some have been a protest against specific policies or government failure in general. Others have been motivated to change cultural norms. The climate movement is global in scope and focuses on the injustice of risks being imposed on future generations.

Government Failures

Of the first we can start with the student riots and occupations that brought France to a standstill in May of 1968. The students' occupations and demonstrations were joined by a brief general strike of workers in major industries. Judt, while describing the events as the greatest movement of social protest in modern France, finds it hard to explain what it was all about. If the students and strikers aimed to bring down the president (Charles de Gaulle) and his Prime Minister George Pompidou, that was fraught with danger, for as the philosopher Raymond Aron pointed out, 'To expel a President elected by universal suffrage is not the same as expelling a king'.[17] The student movement was met with a popular backlash and counter-demonstrations much larger than those of the students' manifestations. The Gaullist parties prevailed.

The movement against the French government of President Emmanuel Macron by the Gilets Jaunes (Yellow Vests) between 2017 and 2019, though also taking politics to the streets, was different in several ways from the 1968 protests. They used violence against property as their weapon, and the political target of their fury was quite specific: the abolition of the wealth tax in the name of competition on European markets 'financed by a carbon tax that fell heavily on the poorer half of the population'.[18] The seething anger broke out again with even more vicious and undiscriminating violence in 2023 triggered by the police killing of a teenager (a French citizen of Algerian descent) after a traffic stop.

Whereas the 1968 movement arose from the protests of students, arguably a fairly privileged middle-class group, the yellow vests the protesters wore signaled their working-class origins. Whereas the student movement was of the ideological left, Piketty considers the Gilets Jaunes to be a 'nativist' (for the French nation) movement driven by disillusionment with the neoliberal ideology of the European Union.[19] The concept of nativism is discussed further in Chapter 11. The movement 'called for redistributive economic policies like a wealth tax, increased pensions, and a higher minimum wage'.[20]

In Britain a powerful social movement formed in opposition to the Thatcher government's poll tax, a flat-rate per capita tax on all real estate to replace the local council tax – the rates – proportional to the value of the property. A socialist group called Militant Tendency, realising what effect this regressive tax change would have on the lowest paid and those without wealth, set in motion a network of groups and unions opposed to the tax. During 1990, over six thousand actions against the tax were held nationwide. Shop windows were smashed and Local Council meetings discussing the tax were invaded. Some demonstrators and police were injured. Demonstrations involving thousands of people were met with violence, with mounted police charging into the crowds.

Both the Conservative Prime Minister Thatcher and the Leader of the Labour Party condemned the movement calling the demonstrators a 'rent-a-mob' and 'toy town revolutionaries'. But the movement escalated. On 31st March a huge demonstration took place in central London with a crowd estimated by police at 200,000 converging on Trafalgar Square. Vehement opposition to the tax was registered nationally, with opinion polls indicating 78 per cent of respondents opposed the tax. After Margaret Thatcher resigned as Prime Minister in 1990 the poll tax was abandoned a year later by her successor as Conservative leader, John Major.

Cultural Norms, Identity, and Human Rights

The second kind of social movement within democracies has focused on racial, cultural and environmental injustice, and human rights. Examples are the Black Lives Matter (BLM) movement against racial prejudice and police violence, the #MeToo movement against sexual violence and abuse, the LGBTQI+ movement for sexual and gender rights, and the Climate Movement.

BLM, beginning in 2013, is a decentralised political and social network aiming to focus attention on racism towards and inequality of the black population in the USA. It is motivated particularly by repeated incidents of the killing of black people by the police. The movement advocates policy changes to increase civil liberties. However, activist Frank Roberts writes that BLM has 'always been more of a human rights movement than a civil liberties movement ... less about changing specific laws than about fighting for a fundamental reordering of society in which Black lives are free from systematic dehumanization'.[21]

The movement gained international attention following the brutal killing of George Floyd by police. The 2023 riots in France also seem to centre on systemic violence and discrimination against non-whites, though this is as yet unclear.

#MeToo, starting in 2006, is a social movement in which digital space is devoted to publicising women's experiences of sexual abuse or harassment, especially in the workplace. The aim of the movement is to empower sexually assaulted people by enabling them to talk about their experiences 'through empathy, solidarity and strength in numbers'.[22]

The women's movement, having won democratic rights, still has a long way to go. In the later twentieth and early twenty-first centuries women have risen to power in governments and some business boardrooms, but in institutions designed by men for rule by men. Critique of the organisational *culture* of institutions is relatively new. In a survey of the field Chappell and Waylen conclude: All institutions are profoundly imbued with gender. This is not always perceived but has important consequences that are both intended and unintended. The operation of gendered rules, norms, and practices (and their intersection with race, class, and sexuality structures) influences institutional design choices and processes.[23]

LGBTQI+ is a loose collection of movements and networks advocating equal rights for people who identify themselves as lesbian, gay, bisexual, transgender, intersex, etc. The movement is about 'gender', that is about 'how a person feels they best align with the socio-behavioural norms or rules of each gender type'.[24] Essentially the movement asserts the right of people, traditionally considered deviant by society, to be who they want to be and to have the same right to that 'being' as everyone else.

Some theorists have identified distinctive traits of contemporary social movements. For instance, they see the new social movements as addressing changes in identity, lifestyle, and culture rather than changes in specific policies. They suggest that the key actors in the movements are more likely to come from the middle class. They tend to operate with loose networks of supporters rather than formal organisations.[25]

However, attempts to classify social movements are problematic. Middle-class activism has grown but not replaced working-class activism. In any case, as mentioned above, the middle class has for many years played a significant role in social change. The Black Lives Matter movement, though originating in the USA, has resonated worldwide in countries such as Australia, Canada, and France where racial injustices continue to occur.

While a protest at all racial injustice, Black Lives Matter comes mostly from injustices inflicted on the poor and working class. Looking back to the nineteenth century, the working-class movements were not lacking in awareness of their identity. A class for itself, as Marx argued, recognised and asserted its special identity. Or perhaps we should say 'identities' – as miners, engineers, building labourers – written into the DNA of their trade unions.

The Climate Movement

The Climate Movement is a global movement to pressure governments and industries to take action on the causes and impacts of climate change. The movement has in recent years involved large-scale protest actions including the People's Climate Marches of 2014 and 2017, and the climate strikes in 2019. 'Youth activism and involvement has played an important part in the evolution of the movement after the growth of the Fridays For Future strikes started by Greta Thunberg in 2019'.[26]

The Climate Movement is more than just a branch of the environmental movement. Whether it knows it yet or not, the Climate Movement must include societal change, and that requires not just change *within* democracies but *change of democracy*. In Greta Thunberg's magnificent collection of short essays by leading scientists and political analysts that theme is beginning to emerge, for example in the chapters by Naomi Oreskes, Bill McKibben, Thunberg herself, and the essay by Stuart Capstick and Lorraine Whitmarsh.[27]

Capstick and Whitmarsh argue that 'Our spheres of influence extend from private and personal choices, through persuading and supporting others, to organizing and agitating for change, and, ultimately becoming a part of remaking the very systems and cultures that make up society'.[28]

Social Movements and the Failure of Democracies

The point of this chapter is not to analyse the precise characteristics of social phenomena, but to rebalance the focus of earlier chapters on state actors and institutions in order to show that movements in civil society have played, and will continue to play, an influential and perhaps central role in social transformations. But there is something different between the social movements *for* democracy and the social movements *within* democracy, and something common to both that is mostly *absent*. That absence is a focus of social movements on democracy itself.

When the core assumptions of the 'Master Narrative' of social democracy began to erode and crumble under the neoliberal transformation, 'they took with them, as Judt notes, not just a handful of public-sector companies but a whole political culture and much else besides'.[29] What was all but destroyed by the neoliberal regression was the power of the state to deliver social progress.

Democracy promised to deliver the state into the hands of the people. But the neoliberal regression, in the interest of the proprietors of national and global wealth, succeeded in reducing the power of the state to little more than a servant of 'the economy'. We might well ask, now that we have won democracy, what have we actually gained in terms of the *ends* of democracy: freedom, justice, equality, and the abolition of poverty?

The power of the state has been eroded internally and externally, nationally and internationally. Struggles for social and environmental protection continue, but they take different forms, emphasising identity and culture. They create new demons to fight against inspired by false conspiracy theories peddled on social media. They say: We have democracy, but our lives don't seem to be improving. So, the problem must lie elsewhere: in immigration, in the 'deep state', in the European Union, in democracy itself.

Protesters sometimes turn to violence. When the democratic state is seen to be powerless to deliver social progress, they seek to overturn the democratic state. Forms of fascism come to be seen as attractive. Today, with the focus on the USA, fears that 'democracy is in trouble' are frequently raised. But such fears need to be viewed in historical context. Democracies *are* mostly flawed, but the civil movement for social and environmental protection identified by Karl Polanyi is alive and well, if sometimes diverted into anti-democratic channels.

Today democratic societies face a number of concatenating crises, a 'polycrisis': a term denoting a single interconnected convergence.[30] The destruction of the Russian polity and nascent democracy has produced war in Europe – irrational but extremely threatening with its global impact on food and energy supplies. Ever since the Great Financial Crisis of 2008, the international economy has tottered from economic crisis to crisis.

The politics of the democracies has been reshaped and destabilised by the so-called 'populists' whose promises to deliver social progress are never realised. The plague of Covid-19 has become endemic with its own economic impacts. The

United Nations Security Council is failing to deliver international security when two of its nuclear-armed members are abusing the UN Charter of Human Rights.

Beyond and behind all these superficial crises the climate crisis develops. That, and its implications for societal transformation and the pressing need for ideological and institutional development of democracy are the subjects of the next part of this book.

Notes

1 https://en.wikipedia.org/wiki/Timeline_of_women%27s_suffrage (accessed 01/12/2022).
2 Herre, B. Ortiz-Ospina, E. and Roser, M. (2013).
3 Judt, T. (2006) p. 559.
4 Ibid.. 567 citing the Hungarian theorist Mihaly Vajda.
5 Ibid. p. 568.
6 Ibid. p. 568
7 Ibid. p. 569.
8 Ibid. p. 571. The work of environmental activists and organisations in the Eastern Bloc is discussed at length in Carter, F.W. and Turnock, D. eds (1993) pp. 237–241.
9 Judt, T. (2005) Chapter XVIII.
10 Ibid. p. 577.
11 Ibid. p. 605.
12 Ibid. p. 606.
13 Ibid. p. 607.
14 Ibid. p. 608.
15 Ibid. p. 610.
16 Ibid. p. 611.
17 Cited by Judt, T. (2005) p. 411.
18 Piketty, T (2020) p. 884.
19 'Nativism' is the term Piketty uses to identify a tendency for social movements to place national interests above global or regional interests, to resist economic globalisation and oppose liberalism.
20 https://en.wikipedia.org/wiki/Yellow_vests_protests (viewed 15/11/2022).
21 Roberts, F. (2018).
22 Kennedy School of Government Case Study (2020) 'Leading with Empathy, Tarana Burke and the Making of the Me Too movement', Cambridge: Harvard University, https://case.hks.harvard.edu/leading-with-empathy-tarana-burke-and-the-making-of -the-me-too-movement/, (accessed 17/11/2022).
23 But see Chappell L. and Waylen, G. (undated).
24 https://news.ucdenver.edu/what-is-the-i-in-lgbtqia/ (accessed 17/11/2022).
25 See Melucci, A. (1980) and Pichardo, N.A. (1997).
26 https://en.wikipedia.org/wiki/Climate_movement (accessed 22/11/2022)
27 Thunberg, G. ed. (2022) See: Oreskes, N. 'Why Didn't They Act' (Chapter 1.7), McKibben, B., 'The Persistence of Fossil Fuels' (Chapter 4.5), Capstick, S. and Whitmarsh, L., 'Individual Action, Social Transformation' (Chapter 5.2).
28 Capstick, S. and Whitmarsh, L. (2022).
29 Judt, T. (2005) p. 559
30 Touve, A. (2021) *ip*. 6: 'A term that has come into use in the last decade. European Commission president Jean-Claude Juncker borrowed the idea from the French theorist of complexity Edgar Morin. Juncker used it to capture the convergence between 2006 and 2016 of the Eurozone crisis, the conflict in Ukraine, the refugee crisis, Brexit, and the Europe-wide upsurge in nationalist populism'.

References

Capstick, S. and Whitmarsh, L. (2022) 'Individual action, social transformation', in Thunberg, G. ed. *The Climate Book*, London and New York: Allen Lane/Penguin/ Random House, pp. 328–330.

Carter, F.W. and Turnock, D. eds. (1993) *Environmental Problems in Eastern Europe*, London and New York: Routledge.

Chappell, L. and Waylen, G. (undated) 'Gender and the hidden life of institutions', prepublication release for *Public Administration* 91/3 pp. 599–615.

Herre, B. Ortiz-Ospina, E. and Roser, M. (2013) 'Democracy'. Published online at *OurWorldInData.org*. https://ourworldindata.org/democracy (accessed 15/11/2022).

Judt, T. (2006) *Post War, A History of Europe Since 1945*, London: Penguin.

Kennedy School of Government Case Study. (2020) *Leading with Empathy, Tarana Burke and the Making of the Me Too Movement*, Cambridge: Harvard University (https://case .hks.harvard.edu/leading-with-empathy-tarana-burke-and-the-making-of-the-me-too -movement/ (accessed 17/11/2022).

Melucci, A. (1980) 'The new social movements: A theoretical approach', *Social Science Information* 19: 199–226.

Pichardo, N.A. (1997) 'New social movements, a critical review', *Annual Review of Sociology* 23: 411–430.

Piketty, T. (2020) *Capital and Ideology*, tr. Arthur Goldhammer, Cambridge: Bellknap Press of Harvard University.

Roberts, F. (2018) 'How Black lives matter changed the way Americans fight for freedom', American Civil Liberties Union. https://www.aclu.org/news/racial-justice/how-black -lives-matter-changed-way-americans-fight (accessed 28/11/2022).

Thunberg, G. ed. (2022) *The Climate Book*, London and New York: Allen Lane/Penguin/ Random House.

Touve, A. (2021) *Shutdown, How Covid Shook the World's Economy*, London: Allen Lane.

PART 2
The Climate Transformation

7
CLIMATE CRISIS

Introduction

The curious proposition by conventional market economists that 'the economy' can be understood as an isolated system with a circular flow of goods and services between firms and households is a pernicious fairy tale that literally *externalises* all of nature.[1]

Even as we talk constantly about the 'climate crisis' or 'environmental crisis', we unwittingly *externalise* the crisis. It is something 'out there' being done to us. So, we have to stop the climate from doing it to us. The reality is exactly the opposite. The biosphere and its climate respond to whatever we throw at it. It is passive, we are active. The biosphere is adapting to us – by heating the air and water, changing the weather, and sometimes by extinguishing species.

The way it is adapting of course rebounds upon us, forcing us to adapt to the changing biosphere. The climate crisis is *first and only* a crisis of our societies, institutions, and ideologies. Those are what we have to *transform* today in order to adapt to and, if possible, mitigate the mess our societies have made.

History does not repeat itself. But the previous chapters have traced a story of transformation of the ways in which democratic societies have organised the collective consumption of nature for human benefit. For that is what an 'economy' is: the organised exploitation and consumption of the natural world for human use. That story, extending the historical narrative of Polanyi, can provide some clues to the present.

Origins and Responses

The origins of today's climate crisis lie in the industrial use of fossil fuels – coal, oil, and gas, to produce energy. The energy released from combustion was used to

DOI: 10.4324/9781003382133-9

drive industrial production, transport and agriculture, global mobility (land, sea, air, and space), and mobile warfare. At the same time, as Polanyi argues, industrial production required an ideological and institutional transformation to create a market-based society.

Yet, as Piketty remarks,

> Although there can be no doubt about the progress made between the eighteenth century and now, there have been phases of regression, during which inequality increased and civilization declined. The Euro-American Enlightenment and the Industrial Revolution coincided with extremely violent systems of property ownership, slavery, and colonialism.[2]

The ultimate regression, however, is in the biosphere. Prosperity and wealth founded on fossil fuels could not continue indefinitely. Burning the carbon sequestered in rocks over millions of years put carbon gases back into the atmosphere causing climate change. Our world has been slower to absorb the science of climate change than the pace of climate change itself.

The scientific understanding of global heating began in the nineteenth century with the study of ice ages and other natural changes in the ancient climate. Many theories were advanced for the causes of these changes, including the action of volcanoes and variations in solar radiation. Among these theories was that of the 'greenhouse effect' driven by the composition of the gases in the atmosphere.

The French mathematician Joseph Fourier was the first to propose (in 1827) that the Earth's average temperature was the result of a balance between incoming energy from the sun and energy radiated back into space. Theoretically, Fourier proposed, the Earth should be much colder than it is. The existence of carbon gases in the atmosphere keeps the earth from becoming an icy wilderness.[3]

In 1896, the Swedish chemist Svante Arrhenius predicted that by releasing the products of burning fossil fuels into the atmosphere, humanity would gradually warm the Earth by several degrees. His findings did not trigger concerns until the 1950s when a few scientists started pointing out that this gradual warming could, in the long term, have catastrophic consequences.[4]

Changes in the carbon content of the Earth's atmosphere have occurred many times over the Earth's long history, but never so rapidly as during the last 150 years. There is a time lag, how long we don't know, between changes in the atmospheric carbon level and the heating effect on the global climate.

From a level of atmospheric carbon dioxide of less than 280 parts per million (ppm) at the beginning of the twentieth century, in 1960 the concentration had risen to around 300 ppm.[5] Over the next 10 years the level rose to 330 ppm. The United Nations held the first world climate conference in 1979. By 1980, concentration had risen to 340 ppm. In 1993, the expert on international law, Daniel Bodansky warned that as a result of increasing greenhouse gases in the atmosphere, global warming could increase 'by 0.2 to 0.5 degrees Celsius per decade, and up to 5 degrees by the end of the 21st Century'.[6]

The Kyoto Protocol entered into force in 2005, by which time the atmospheric carbon concentration had risen to 380 ppm. Since the first IPCC Assessment Report there have been 28 Conferences of the Parties to the Framework Convention (COPS), each report of the conference more strident than the last in warning of the dangers of the heating world. The Kyoto Protocol, the Copenhagen Accord, and the Paris Agreement have all been concluded with the ambition of reducing atmospheric carbon.

Yet the US National Oceanic and Atmospheric Administration (NOAA) reported that by 2022 atmospheric carbon concentration had reached 417.2 ppm. Carbon gases in the atmosphere have continued to grow rapidly in a uniform upward curve.[7] The growth of atmospheric carbon shows no sign of abating. Beyond 3° of global heating the results begin to threaten human civilisation as well as many of the species of animals on Earth (see Chapter 1).[8]

Mark Lynas observes,

It takes thousands of years for warmer temperatures to penetrate into the darkest depths of the ocean for example; and for as long as the seas carry on warming, the atmosphere cannot reach equilibrium, because heat is still transferring downward. This is an example of the planet's 'thermal inertia'. Temperatures will always lag behind changes in 'forcing' from solar radiation or greenhouse gases, because of the long response time of the Earth system.[9]

In 2000, a group of scientists from the UK Hadley Centre reported modelling which showed that 'carbon-cycle feedbacks could significantly accelerate climate change over the twenty-first century'. They reported that: 'under a 'business as usual' scenario, the terrestrial biosphere acts as an overall carbon sink until about 2050 but turns into a source thereafter'.[10] In other words, major sources of carbon storage such as the Amazon rainforest after about 2050 become carbon emitters, resulting in a 3° rise in global temperatures as early as 2050 (in just 17 years' time). A study published in *Nature* in 2023 builds on the Hadley Centre findings and later research.[11] The study found that key ecosystems (including the Amazon rainforest) could collapse sooner than previously believed. The study shows that 'tipping points' driven by global heating can, working together, amplify and accelerate each other.[12]

The world is not moving away from a fossil fuel-based economy fast enough to prevent the rise of global average temperature past 2° of warming. The fossil-fuelled economy persists for three main reasons. First because, as McKibben puts it, fossil fuels 'have produced the world we know'. We live in a world 'pulsing with energy'. A single barrel of oil can do the work of 25,000 hours of human labour. Fossil fuels have enabled enormous prosperity. Secondly, because the inertia built into the industrial system of production and consumption takes time to change and adapt to the need to not emit carbon into the atmosphere. Thirdly, because vested interests have worked to delay the process of global change.[13]

Not only climate scientists and environmentalists are warning of climate catastrophe. Reports from the security services in the USA, UK, and Australia are also issuing warnings. Not much media attention is given to these sources, and the services themselves are secretive in function. Yet the warnings should shock us. In 2023, the former head of the Australian Defence Force, Admiral Chris Barry, now a member of the Australian Security Leaders' Climate Group, went public calling for the latest Australian report to be published. He did not mince his words, 'We are worried about the possible collapse of societies because of starvation, a lack of fresh water and shortages of food supplies'. The threat from climate change, he said, poses a bigger risk than the threat from China, 'Only a nuclear war could be more catastrophic'.[14]

A briefing paper from the White House in 2021 states:

The three broad categories of risks are: 1) increased geopolitical tension as countries argue over who should be doing more, and how quickly, and compete in the ensuing energy transition; 2) cross-border geopolitical flashpoints from the physical effects of climate change as countries take steps to secure their interests; and 3) climate effects straining country-level stability in select countries and regions of concern.[15]

Research led by Professor Tim Lenton of the University of Exeter, published in *Nature Sustainability*, has found that billions of people could be forced out of their 'human climate niche' as the planet warms. 'Up to 1 billion people could choose to migrate to cooler places, the scientists said, although those areas remaining within the climate niche would still experience more frequent heatwaves and droughts'.[16]

Environmental journalist Abrahm Lustgarten reports,

Around the world, rising temperatures and climatic calamity are unsettling ever larger numbers of people. As droughts, floods, storms and heat make it difficult to farm, work, and raise children, populations are moving in search of temperate conditions, safety and economic opportunity. Food security is fast becoming the planet's most significant human threat, leading the world to the precipice of a great climate migration.[17]

Lustgarten provides vivid examples of the beginning of migration from Central America: 'Half a million people across El Salvador, Guatemala and Honduras faced immediate and acute malnutrition – even starvation – as farmers struggled to produce food'.[18] The droughts were driven by repeated La Niña cycles of the Southern Oscillation weather system.

In the absence of radical intervention by governments, every decade will take us closer to a 3° rise, and every decade will contain more and more frequent catastrophic events: droughts, floods, overheated cities, cyclones and tornados, firestorms over enormous areas and rising sea levels threatening small islands and coastal cities.

The Changing Weather and Its Costs

On the way to the Neo-Pliocene, climate change has started to drive changes in the local weather across the whole planet. This was evident in 2008 and has become clearer in the 15 years since Lynas's book was published. Exactly how and where a heating world affects local weather is a complex problem. The focus now is on 'weather systems' including especially the El Niño Southern Oscillation (ENSO) affecting countries on both sides of the Pacific Ocean such as Australia and Indonesia in the East, and Chile and Peru in the West. The ENSO blog tells us:

> Though ENSO is a single climate phenomenon, it has three states, or phases, it can be in. The two opposite phases, "El Niño" and "La Niña," require certain changes in both the ocean and the atmosphere because ENSO is a coupled climate phenomenon. "Neutral" is in the middle of the continuum.[19]

The El Niño phase causes floods in Peru's Atacama Desert and droughts in Indonesia and Australia – 'it has ripple effects right across the planet' including the Indian Monsoon and hurricane formation in the Atlantic.[20] Whether the oscillation will become stronger or weaker or whether the El Niño phase will become permanent (and cease to oscillate, as seems to have occurred during the Pliocene) is the subject of an ongoing debate. Australian meteorological studies now include three other weather systems: the Indian Ocean Dipole, the Madden-Julian Oscillation, and the Southern Annular Mode. Climate variability is now known to be driven 'by many significant climate features that will have varying levels of impact in different regions at different times'.[21]

In 2020, flooding throughout South Asia (Afghanistan, India, Bangladesh, Nepal, Pakistan, and Sri Lanka) caused an estimated US$105 billion damage and 6,511 deaths across the region – the costliest stand-alone flood in modern history.[22] According to a study by the UK charity *Christian Aid*, in July 2021 floods in China's Henan Province killed more than 300 people and inflicted US$17 billion damage. Floods in Europe in August caused 240 deaths and economic costs of US$43 billion. A month later Hurricane Ida made landfall on the east coast of the USA killing 95 people and causing US$65 billion damage.[23]

A year later, from June to October 2022, massive flooding in Pakistan left over 1,700 people dead, 12,000 injured, and 2.1 million homeless. The floods covered at least one-tenth of the country.[24] These deadly floods were the worst in the country's history and caused an estimated US$14.9 billion damage and US$15.2 billion of economic losses. The floods came from heavier-than-normal monsoon rainfall and melting glaciers that followed a severe heatwave.

The Christian Aid study for 2022 assessed the cost of economic damage of the top ten extreme weather events. 'The ten most financially costly events all had an impact of $3 billion or more'. Hurricane Ian struck Cuba in September, costing US$100 billion and displacing 40,000 people. The drought and heatwave in Europe cost US$20 billion. The study states:

> While the report focuses on financial costs, which are usually higher in richer countries because they have higher property values and can afford insurance, some of the most devastating extreme weather events in 2022 hit poorer nations, which have contributed little to causing the climate crisis and have the fewest buffers with which to withstand shocks.[25]

In many cases both flooding and fires have had disastrous effects for natural ecosystems and biodiversity. For example, in Eastern Australia, the massive bushfires of 2019–2020, following years of drought and the hottest summer on record (El Niño phase), burned through an estimated 243,000 square kilometres of forest, extending from the state of Queensland in the north to those of Victoria and South Australia in the south and west. At the peak of the fires, air quality across the area dropped to hazardous levels, with smoke haze blanketing Sydney and Melbourne.[26] A similar fire state occurred in North America in 2023 with similar results. The Australian fires of 2019 were followed by a triple occurrence of the La Niña phase in 2021–2022 which brought devastating floods.

Air pollution from burning carbon is also a critical problem. Worldwide, 8.7 million people die every year as a result of air pollution – 'more than AIDS, malaria and tuberculosis combined'.[27] Stopping the burning of fossil fuels yields big and immediate advantages in reducing air pollution as was found in London after the great smog of 1952, following which burning coal for heating was eventually banned. Today, indoor air pollution kills an estimated 4 million people annually, mostly in poor countries that still use wood or other carbon sources for cooking fuel.[28]

These are just a few examples of how climate change is driving extreme weather which is harming people and nature. Today, extreme weather events are reported almost daily. Such events are becoming 'normal' and quickly forgotten in the daily news cycle. Meanwhile, longer-term changes are occurring. Glaciers are melting, reducing water supplies for dependent populations. The ice sheets in Greenland and Antarctica are melting. Sea levels are rising threatening coastal communities. The Arctic Ocean is losing its ice cover, the Amazon basin is drying out. The human exploitation of the world's environment by societies so 'successfully' locked into the logic of market forces is relentless and causing degradation and pollution.

A peer-reviewed study by the Natural Resources Defence Council for the USA in 2008 concluded for the USA, that if present trends continue, the total cost of global warming will be as high as 3.6 per cent of the US Gross National Product: 'Four global warming impacts alone – hurricane damage, real estate losses, energy costs, and water costs – will come with a price tag of 1.8 per cent of U.S. GDP, or almost $1.9 trillion annually (in today's dollars) by 2100'.[29]

The authors state:

> The sad irony is that while richer countries like the United States are responsible for much greater per person greenhouse gas emissions, many of the poorest countries around the world will experience damages that are much larger as a percentage of their national output. They also acknowledge that it is difficult to put a price on many of the costs of climate change, 'loss of human lives and health, species extinction, loss of unique ecosystems, increased social conflict, and other impacts extend far beyond any monetary measure'.[30]

In 2022, the science and environment writer Eugene Linden reported a study by Moody Analytics that the global economic toll of 2° of warming would be US$69 trillion. Linden writes, 'Three degrees of warming would produce a world that hasn't existed since humans emerged as a species. There was plenty of life back then but no humans. It's certain that such a world could not support 7.8 billion people'.[31]

As global heating and its effect on local weather continue, the kind of costs that the world will face can be divided into three categories; pre-emptive costs, restorative costs, and adaptive costs.

Pre-emptive Costs

These are the costs associated with what is usually called 'mitigation'. They are costs incurred in efforts to reduce carbon emissions and otherwise reduce atmospheric heating. They include conversion of power generation to renewable sources such as wind and solar backed up by arrays of large-scale batteries, wave or tidal energy from the sea, geothermal energy, 'green' hydrogen converted from water using renewable energy, traditional hydroelectric generation and 'pumped hydro' in which water stored in high-level dams can be released to generate electricity in times of need, and can then be pumped uphill back into the dams using renewable energy.

All these mitigation technologies are expensive and often encounter major hurdles in implementation. In Australia, the Snowy pumped hydro scheme using two existing dams in the Snowy mountains has run into long delays and cost blow-outs while boring a new tunnel through soft ground.[32] Even more significant is the cost of refitting the existing energy grid (and market) in the vast territory of Eastern Australia to receive and distribute energy from wind and solar, now estimated at Aus$5.2 billion and almost certain to rise.[33] The cost problem is exacerbated by the short time horizon left to make the transition to renewables. Glen Peters warns, 'The amount of time to make the energy transition is vanishing. To achieve it, we need all the tools in the toolbox'.[34]

Further pre-emptive costs are incurred from electrification of transport systems, reduction of land clearing for agriculture, and preservation of existing forests and planting of new ones. But, as Karl-Heinz Erb and Simone Gingrich observe, rather than planting new forests to be harvested for wood, 'protecting forest carbon sinks by reducing wood harvesting appears to be the optimum strategy: forest carbon sinks sequester atmospheric carbon of 10.6 $GtCO_2$ (gigatons) equivalent per year, compensating for around 30 per cent of total annual emissions'.[35] Reduction of land clearing and forest sequestration are not costless because of the value foregone by landowners who might otherwise benefit economically from logging – and who therefore need to be compensated.

Climate geo-engineering is also on the horizon. There are serious doubts about the risks and reversibility of most geo-engineering concepts such as sun-dimming injection of aerosols into the atmosphere.[36] But one approach is both safe and reversible: carbon-dioxide removal from the atmosphere. There are two main approaches: biological and mechanical. Biological approaches include increased carbon storage in agriculture, forestry, and other land use sectors, such as by afforestation, reforestation, and various innovations to increase soil carbon.[37] The mechanical approach either captures carbon emissions from power stations and other sources (carbon capture and storage), or directly removes carbon gases from the air (direct air capture). Both require captured carbon to be sequestered underground or in the ocean.

A direct air capture plant is already operating in Iceland. But the costs of scaling up such plants are likely to be prohibitively vast, and sequestration creates major ecological problems. The environmental engineers Patrick Moriarty and Damon Honnery in their study of the future of energy conclude that neither biological nor mechanical methods for CO_2 removal – in the increasingly unlikely event that they are implemented at large scale – can ever be more than temporary solutions for our climate change-related energy problems.[38]

Restorative Costs

Much less attention has been given to the rising costs associated with rescuing or helping out those people and ecosystems afflicted by extreme weather events in various ways. Rich countries can afford to provide emergency assistance to businesses and householders afflicted by floods and bushfires. But the assistance rarely covers the full costs, and much restorative work is left to volunteer organisations.

The assumption has been that it is the responsibility of individuals to take out insurance against such contingencies. It is becoming clear, however, that private insurance cannot protect against climate-driven risks. The costs are simply much too great. The fires and floods in Australia from 2018 to 2022 have made the costs of insurance unaffordable. By 2030, due to worsening extreme weather events, about 1 in 25 properties will be uninsurable.[39] A study by the Rand Corporation found that wildfire risks with increasing greenhouse emissions in California would 'challenge' the insurance market in terms of policy affordability, coverage adequacy, and insurer profitability.[40]

Some of the most-favoured sites for housing in rich countries are beside coasts and rivers. Some of these areas, with beautiful sea views and beaches, and among tree-lined rivers, will become uninhabitable and uninsurable due to rising seas and flooding rivers.

The problem becomes much greater at the international level. The 'aid' available from UN organisations, competing with a range of other emergencies (such as earthquakes, famines, and wars), cannot possibly provide more than temporary and minimal assistance to flood victims in places such as Pakistan and India, whose governments already face enormous problems of poverty and social insecurity.

In the absence of life support in impoverished regions suffering from conditions resulting from global heating, the only possibility will be for people to leave. The issue of social justice for 'environmental refugees' was recognised in 2001. In this analysis, Adrianna Semmens identified 'environmental refugees' as those who have been coercively uprooted from their social-environmental domiciles by the imperative of commercial development. She concluded that, in order to maximise justice for environmental refugees, 'it would seem that a transnational institution or regime is needed beyond the institutions already formed for refugees'.[41]

Adaptive Costs

Finally, human habitation almost everywhere will have to adapt to the changing weather. When floods, fires, or coastal erosion make some areas of settlement uninhabitable, new housing will have to be located in safe areas to accommodate those driven out.

Poorly insulated houses will need to be refitted with high-quality insulation and, probably, air conditioning. Cities will need their streets to be shaded, with tree cover to limit the heat-island effect. Urban heat islands occur in cities with concentrations of pavement, buildings, and other surfaces that absorb and retain heat. In order to meet the costs of adaptation, there will be a need for strong and far-sighted town and country planning armed with powers equivalent to those used to build the British new towns after the devastation of the Second World War. If dislocated populations in poorer countries are to be provided for within national borders (and mass migrations avoided), the need for international aid will dramatically increase.

Denial, Opposition and the Persistence of Fossil Fuels

The world has had 30 years to respond to the call from the United Nations to act on climate change. The science of climate change has been known for much longer. McKibben reminds us that vested interests are holding back the transition to cheaper and readily available sources of energy. The power of fossil fuel corporations, and their ability to influence policymakers, have put a brake on the development of cheaper, cleaner energy: 'the fossil fuel industry has single-mind-edly used its power to delay action'.[42] The biggest political donors in American history, McKibben writes, were also the biggest oil and gas barons.

Stanford PhD candidate Benjamin Franta spent years seeking out the hidden history of climate change in American historical archives. In 2021, he reported his findings from the transcript of a 1959 petroleum conference in Delaware. At the conference, Franta tells us, Edward Teller, the scientist who helped invent the hydrogen bomb, warned of the danger of climate change. Teller explained to the conference that whenever you burn conventional fuel, you create carbon dioxide. Its presence in the atmosphere causes a greenhouse effect. If the world kept using fossil fuels, the ice caps would begin to melt, raising sea levels. Eventually, all the coastal cities would be drowned.[43]

Franta reported that in 1965 the president of the American Petroleum Institute, Frank Ikard, spoke about a report by President Johnson's scientific advisers. Accordingto Franta, Ikard said that:

One of the most important predictions of the report is that carbon dioxide is being added to the earth's atmosphere by the burning of coal, oil, and natural gas at such a rate that by the year 2000 the heat balance will be so modified as possibly to cause marked changes in climate.[44]

In the 1970s, the API formed a secret task force, including many of the major oil companies, to monitor climate change science. In 1980, the task force invited the Stanford University scientist John Laurmann to report on the state of climate science. Laurmann warned that a rise of 2.5° by 2038 would have 'major economic consequences'. A rise of 5° by the 2060s would have 'globally catastrophic effects'. That same year, Franta reports, the API called on governments to triple coal production worldwide, insisting that there would be no negative consequences.

Exxon also had a secret research programme, Franta reports. In 1982, Exxon produced a comprehensive internal report which predicted almost exactly the amount of global warming we have seen today, and in the long run warming that would 'produce effects that would be catastrophic (at least for a substantial fraction of the Earth's population)'. Other oil companies such as Shell and Total knew about the effects of climate change. A confidential report by Shell predicted the greatest change in recorded history, including destructive floods, abandonment of entire countries, and even forced migration around the world.

Franta's research confirms the investigative work of Naomi Oreskes.[45] Oil companies, instead of turning their operations towards renewable energy, conducted a campaign to deliberately create confusion about climate science. Oreskes follows up in 2022, 'Through advertisements, public relations campaigns, reports commissioned from "experts for hire", and more, the carbon-combustion complex deliberately created confusion about the climate crisis'. She continues:

Many of the strategies and tactics were taken directly from the tobacco industry, including cherry-picking and misrepresenting scientific evidence; promoting outlier scientists to create the impression of scientific debate where there was little or none; funding research intended to deflect attention from the primary causes of climate change and falsely portraying the fossil fuel industry as supporting 'sound science' rather than protecting profits.[46]

Companies also deflected attention from their role 'by insisting citizens should take "personal responsibility" by lowering their carbon footprints'.[47] Herein lies the link with ideology. At the core of the neoliberal regression is the insistence that governments should intervene as little as possible in the interactions of individuals, regulated only by markets and the enshrined ownership of property. Whether in the use of tobacco, asbestos, or fossil fuels the responsibility of production companies goes no further than their responsibility to shareholders to deliver dividends and protect the share price.

To apportion blame, it is fruitless to appeal to companies to behave ethically when their very being is governed by market relations. Everyone surely knows by now that pure liberalism, reducing humans and nature to fictional commodities, does not serve the public interest. Adam Smith himself knew that, as did Karl Polanyi. We repeatedly turn to governments to solve the problems created by market relations: in times of economic depression, pandemic, and in times of climate crisis. In doing so we not only save lives and environments, but we save the 'economy' by creating new pathways for economic growth in, for example, the 'caring economy', renewable energy, electrification, and environmental conservation.

Climate Change and Social Transformation

Throughout Thunberg's anthology of science and philosophy of climate change, we find reference to the way the structures, institutions, and processes of contemporary human societies are interwoven with the transformation of nature – and especially climate change of the Earth's biosphere.

We have seen how Oreskes and McKibben have drawn attention to the failure of honesty and transparency in a market society – backed by the investigations of Franta. Kevin Anderson, reporting on COP 27 in Glasgow, writes of the 'new denialism' of 'expedient technical optimism, of 'negative emissions', and, today, of 'net zero but not in my term of office'. Since the IPCC's first report in 1990, he writes, 'we've dumped more carbon dioxide in the atmosphere than throughout all of human history prior to 1990'.[48]

In recent years, we have seen the blossoming of money-making schemes of emissions trading, carbon offsets, and 'carbon credits' in which intermediary firms offer to negotiate with landowners (often in poor countries) to prevent land clearing and logging of native forests. But regulation of such schemes to enforce compliance is weak or non-existent.[49]

The most egregious denial, however, is of the injustice of climate mitigation based on national inventories. Even under conditions in which richer nations are expected to cut their emissions faster than poorer nations (as agreed at the Rio Earth Summit), 'an individual's average annual emissions, from now to the point of zero global emissions, would still be greater for citizens of high-emitting nations than for those of low-emitting poorer nations'.[50]

Sharing responsibility amongst individual emitters remains a taboo subject. 'For all of us' in rich countries, Anderson writes,

> living within a fair carbon ration would entail profound changes in how 'we' live our lives: the size and number of our houses; how often we fly and in which class; how big and how many cars we have and how we drive them, how many foreign meetings and international conferences we attend and how frequent our field trips.[51]

Climate injustice, as carbon emissions force weather changes across the world, extends to differences between rich and poor in the most basic needs: public health provision, air pollution, food production, and nutrition. As Director General of the World Health Organisation Tedros Ghebreysus writes, 'With the poorest people largely uninsured, health shocks and stresses already push around 100 million people into poverty every year, with the impacts of climate change worsening this change'.[52]

To address social justice in an age of climate change requires transformational change. There are possibilities for transformational action, for instance, as Sunita Narain argues, through pouring investment into meeting the energy needs of the poorest in the world without adding to emissions. The focus must be on those 'who are without the basic infrastructure of electricity to power their homes or to cook their food'.[53]

Narain retains faith in market approaches such as carbon trading and offsets, as do economists such as Nicholas Stern, who observes however, 'Unfortunately, much of the economic analysis of climate change has failed to recognise the necessary urgency and scale of action'. First, by underestimating the immense scale of risks of climate change; secondly by underestimating the potential of alternative sources of energy; and thirdly, 'It has grossly undervalued our descendants' lives through a misleading and ill-founded approach to discounting: we have discriminated against future generations based on their date of birth' (discounting future against present value).[54]

Lucas Chancel and Thomas Piketty point to the vicious circle of climate inequality found both internationally and nationally. Poorer households, which are low CO_2 emitters, rightly anticipate that climate policies will limit their purchasing power. In return, policymakers fear a political backlash should they demand faster climate action. This vicious circle applies especially in democratic countries, but also in non-democratic countries (such as China) that nevertheless depend for their legitimacy and authority on a prospering population at all social levels. There are solutions to this dilemma, but they require deep institutional and ideological change of which Piketty and his co-workers write, and to which we will return in the chapters to follow.

Margaret Attwood, the visionary novelist, puts her finger on the ideological problem. To survive the coming climate catastrophe, she writes, we will need four elements: knowledge, equipment, will power, and luck. What we do not have any more, she writes, are 'literary utopias'. The twentieth century saw two utopian projects crumble, Marxist-Leninist communism of the USSR and National Socialism of Germany's Third Reich. But we also saw Democratic Socialism rise in Britain, Scandinavia, and even eventually in Germany. These were not utopian dreams but practical, ideologically driven modifications to democratic institutions: practical utopias. Attwood asks the key question: 'In a nutshell, could we create a society that sequestered more carbon than it produced, while also creating a fairer, more equal society?' In September 2022, Attwood used a learning platform called 'Disco' for people to interactively create a better future.[55]

There is a sense of unease, even distress, in all these discussions around the climate crisis that the world's political institutions are not capable of managing and adapting to the massive environmental change which is coming. The neoliberal regression which now conditions political responses has produced a level of injustice between those whose activities generate climate change through high levels of carbon emissions and those who suffer its worst effects and are least protected from the changing weather.

As noted in Chapter 1, the transformation we now need must change the fundamental attributes of our socio-ecological system.[56] It is the aim of this book to start answering questions implied in the statement. 1. What is the transformation required – from what present condition to what improved future condition? 2. How might that transformation occur? 3. With what actors and within what time frame?

In the first part of this book, with the help of the work of Karl Polanyi, we traced three specific social transformations, from raw capitalism to democratic socialism, from democratic socialism to neoliberalism, and from communism to liberal capitalism. All of these were driven by dramatic changes in ideology and enacted sometimes under the pressure of popular movements and sometimes by governments and private elites.

Ideology is powerful. It can bind social movements and shape public policy. Both communism and Nazism were powerful ideologies, and wrong – both ethically and practically. The world today desperately needs a new ideology, a new 'practical utopia', but it will not come about through separate individual explorations on internet platforms – though such activities may well inform more collective ambitions.

A practical utopia has to resolve what can be described as 'antinomies', law-like features of society in which two institutional ideals cannot both be met perfectly at the same time. The classic example is that described by Polanyi: a free market and protection of humans and nature (and indeed *between* humans and nature). The best that can be achieved is a trade-off between the two ideals. Trade-offs are variable depending on circumstances, conditional on different nations' political traditions, and always open to revision. There are many possible versions of trade-off. Another kind of trade-off is between decisiveness about policy (exemplified by the majoritarian principle) and the involvement and consultation with the plurality of beliefs, desires, and priorities held by different groups in society.[57]

Before moving to an examination of the elements of 'participatory socialism' in Chapter 11, we need to examine in greater depth two aspects of the groundwork for such an ideology: the state of democracy (Chapter 9) and the state of inequality in the world today (Chapter 10).

Notes

1 Herman Daly exposed this fallacy 30 years ago: 'A search through the indexes of three leading textbooks in macroeconomics reveals no entries under any of the following subjects: *environment, natural resources, pollution, depletion'*. Daly, H. (1992) p. 46.
2 Piketty, T. (2020) p. 19.
3 Though the changing balance of terrestrial temperature is much more complex than the simple image of the greenhouse suggests.
4 Oppenheimer, M. (2022).
5 The following figures for recent atmospheric carbon concentrations are taken from Oppenheimer, M. (2022) Figure 1, p. 28.
6 Bodansky, D. (1993) p. 453.
7 National Oceanic and Atmospheric Administration (NOAA) Research News (2022) *Global Atmospheric Carbon Dioxide Levels Continue to Rise,*)https://research.noaa .gov/article/ArtMID/587/ArticleID/2914/No-sign-of-significant-decrease-in-global -CO2-emissions#:~:text=The%20publication%2C%20produced%20by%20an,percent %20above%20pre%2Dindustrial%20levels) accessed 31/01/2023. The series of UN conferences is contrasted with the unrelenting pattern of atmospheric carbon growth by Oppenheimer, M. (2022).
8 Lynas, M. (2008).
9 Ibid. p. 133.
10 Cox, P.M., Betts, R.A., Jones, C.D., Spall, S.A. and Totterdell, I.J. (2000). Cited by Lynas (2008) p. 139.
11 Ritchie, P.D.L., Clarke, J.J., Cox, P.M. and Huntingford, C. (2021).
12 Willcock, S., Cooper G.S., Addy, J. and Dearing, J.A. (2023).
13 McKibben, B. (2022) 'The Persistence of Fossil Fuels', in Thunberg, G. ed. (2022) p. 219.
14 Knott, M. (2023) 'Push for release of climate risk report', *The Age*, Melbourne, 05/04/2023. (Knott is the National Security Correspondent for the masthead). See also UK Parliament (2023): *Readiness for storms ahead? Critical national infrastructure in an age of climate change.* In the USA (2022): *Climate Change Risks to National Security*, US Government Accountability Office.
15 https://www.whitehouse.gov/briefing-room/statements-releases/2021/10/21/fact-sheet -prioritizing-climate-in-foreign-policy-and-national-security/ (accessed 06/04/2023).
16 Lenton, T, Xu C., Abrams J.F. et al. (2023) reported by Carrinton D. (2023) 'Global heating will push billions outside human climate niche', *The Guardian* 23/05/2023.
17 Lustgarten, A. (2022) 'Climate Refugees', in Thunberg, G. ed. (2022) p. 166.
18 Ibid.

19 https://www.climate.gov/news-features/blogs/enso/what-el-ni%C3%B1o%E2%80%93southern-oscillation-enso-nutshell (accessed 02/02/2023) The ENSO blog is written, edited, and moderated by Michelle L'Heureux (NOAA Climate Prediction Center), Emily Becker (University of Miami/CIMAS), Nat Johnson (NOAA Geophysical Fluid Dynamics Laboratory), and Tom DiLiberto and Rebecca Lindsey (contractors to NOAA Climate Program Office), with periodic guest contributors.

20 Lynas, M. (2008) p.135.

21 https://www.climatechangeinaustralia.gov.au/en/overview/climate-system/australian-climate-influences. (accessed 02/02/2023).

22 https://en.wikipedia.org/wiki/2020_South_Asian_floods (accessed 02/02/2023). Figures from the National Disaster Management Agency of Pakistan.

23 Shepherd, D. (2021) 'Ten worst disasters cost US$235 dollars', Melbourne: *The Age*, 28/12/2021, p. 18.

24 https://en.wikipedia.org/wiki/2022_Pakistan_floods#cite_note-NDMA-3 (accessed 02/02/2023).

25 https://mediacentre.christianaid.org.uk/new-report-top-10-climate-disasters-cost-the-world-billions-in-2022-10035/# (accessed 06/02/2022).

26 https://en.wikipedia.org/wiki/2019%E2%80%9320_Australian_bushfire_season (accessed 06/02/2023).

27 McKibben, B. (2022) 'The Persistence of Fossil Fuels', in Thunberg, G. ed. (2022) Chapter 4.5, p. 219.

28 Shindell, D. (2022) 'Air Pollution' in Thunberg, G. ed. (2022) Chapter 3.4 p. 140.

29 Ackerman, F. and Stanton E.A. (2008). The authors are based at the Global Development and Environment Institute and Stockholm Environment Institute-US Center, Tufts University (https://www.nrdc.org/).

30 Ibid.

31 Linden, E. (2022) p 192.

32 Woodley, T. (2023) *Disastrous tunnelling delays underline folly of Snowy Hydro 2.0 pumped hydro scheme.* 'The latest revelation from the hapless Snowy 2.0 pumped hydro battery project is the staggeringly slow progress of all three tunnel boring machines (TBMs), resulting in extensive delays and cost blowouts'. https://reneweconomy.com.au/disastrous-tunnelling-delays-underline-complete-folly-of-snowy-2-0-pumped-hydro-scheme/ (accessed 15/02/2023).

33 Fernyhough, T. (2022).

34 Peters, G. (2022) p. 227.

35 Erb, K-H and Gingrich S. (2022) p 232.

36 Hällström, N., Stephens, J.C. and Stoddard, I (2022) pp. 233-234

37 Moriarty, P. and Honnery, D. (2022).

38 Ibid.

39 O'Malley, N. (2022) p.3.

40 Dixon, L., Tsang, F. and Fitts, G. (2023).

41 Semmens, A. (2001) p. 86.

42 McKibben B. (2022) p. 221.

43 Franta, B. (2021) 'What Big Oil knew about climate change, in its own words', *The Conversation* 28/10/2021, https://theconversation.com/what-big-oil-knew-about-climate-change-in-its-own-words-170642 (accessed 07/02/2023).

44 Franta, B (2021) Ibid.

45 Oreskes, N. (2010).

46 Oreskes, N. (2022).

47 Ibid.

48 Anderson, K. (2022) p. 205.

49 Australian Broadcasting Corporations (2023) *Carbon Colonialism*, 'Four Corners' report aired 13/02/2023.

50 Anderson, K. (2022) p. 207.

51 Ibid. p. 208.
52 Ghebreyesus T. A. (2022) p. 135.
53 Narain, S. (2022) p. 309.
54 Stern, N. (2022) p. 307.
55 Attwood, A. (2022) p. 362. (see also https://www.discostudios.com/learn-live-with -margaret-atwood-course, accessed 14/02/2023).
56 International Panel on Climate Change (2022).
57 See March A. and Low N.P. (2004).

References

Ackerman, F. and Stanton E.A. (2008) *The Cost of Climate Change, What We'll Pay If Global Warming Continues Unchecked*, Natural Resources Defence Council, New York.

Anderson, K. (2022) 'The new denialism', in Thunberg, G. ed. *The Climate Book*, London and New York: Allen Lane/Penguin/Random House, Chapter 4.2 .

Attwood, A. (2022) 'Practical utopias', in Thunberg, G. ed. *TheClimate Book*, Chapter 5.9.

Bodansky, D. (1993) 'The United Nations framework convention on climate change: A commentary', *Yale Journal of International Law* 14(451): 453.

Carrinton, D. (2023) 'Global heating will push billions outside human climate niche', *The Guardian* 23/05/2023.

Cox, P.M., Betts, R.A., Jones, C.D., Spall, S.A. and Totterdell, I.J. (2000) 'Acceleration of global warming due to carbon-cycle feedbacks in a coupled climate model', *Nature*, 408: 184–187.

Daly, H. (1992) *Beyond Growth, The Economics of Sustainable Development*, Boston: Beacon Press.

Dixon, L., Tsang, F. and Fitts, G. (2023) 'California wildfires can insurance markets handle the risk?' https://www.rand.org/pubs/research_briefs/RBA635-1.html (accessed 06/02/2023).

Erb, K.-H. and Gingrich, S. (2022) 'How can forests help us', in Thunberg, G. ed. *The Climate Book*, Chapter 4.7.

Fernyhough, T. (2022) 'Australia to Spend $5.2 billion on renewable-power grid upgrade' https://www.bloomberg.com/news/articles/2022-12-20 (accessed 15/02/2023).

Franta, B. (2021) 'What big oil knew about climate change, in its own words', *The Conversation* 28/10/2021, https://theconversation.com/what-big-oil-knew-about -climate-change-in-its-own-words-170642 (accessed 07/02/2023).

Ghebreyesus, T. A. (2022) 'Health and climate', in Thunberg, G. ed. *The Climate Book*, Chapter 3.2.

Hällström, N., Stephens, J.C. and Stoddard, I. (2022) 'What about geoengineering?' in Thunberg, G. ed. *The Climate Book*, Chapter 4.8.

International Panel on Climate Change. (2022) *Climate Change 2022, Impacts, Adaptation and Vulnerability, Summary for Policy Makers*, p. 5. Working Group 2 United Nations.

Knott, M. (2023) 'Push for release of climate risk report', *The Age*, Melbourne, 05/04/2023.

Lenton, T, Xu, C., Abrams J.F. et al. (2023) 'Quantifying the human cost of global warming', *Nature Sustainability*, https://doi.org/10.1038/s41893-023-01132-6 (accessed 06/06/2023).

Linden, E. (2022) 'The true cost of climate change', in Thunberg, G. ed. *The Climate Book*, Penguin/Random House, Chapter 3.20, p 192.

Lustgarten, A. (2022) 'Climate refugees', in Thunberg, G. ed. *The Climate Book*, Penguin/ Random House, p. 166.

Lynas, M. (2008) *Six Degrees, Our Future on a Hotter Planet*, Washington: National Geographic.

March, A. and Low, N.P. (2004) 'Knowing and steering: Mediatization, planning and democracy in Victoria, Australia', *Planning Theory* 3/1 pp. 41–70.

Moriarty, P. and Honnery, D. (2022) *Switching Off, Meeting Our Energy Needs in a Carbon Constrained Future*, London and Berlin: Springer Nature.

McKibben, B. (2022) 'The persistence of fossil fuels', in Thunberg, G. ed. (2022) *The Climate Book*, Penguin/Random House, Chapter 4.5, p. 219.

Narain, S. (2022) 'Equity', in Thunberg, G. ed. *The Climate Book*, Chapter 4.25.

National Oceanic and Atmospheric Administration (NOAA) Research News. (2022) *Global Atmospheric Carbon Dioxide Levels Continue to Rise*. https://research.noaa .gov/article/ArtMID/587/ArticleID/2914/No-sign-of-significant-decrease-in-global -CO2-emissions#:~:text=The%20publication%2C%20produced%20by%20an,percent %20above%20pre%2Dindustrial%20levels (accessed 31/01/2023).

O'Malley, N. (2022) 'Uninsurable homes to rise from floods, bushfires', *The Age*, Melbourne 03/05/2022 p. 3.

Oppenheimer, M. (2022) 'The discovery of climate change', in Thunberg, T. ed. *The Climate Book*, Allan Lane/Penguin/Random House, Chapter 1.6, p. 23.

Oreskes, N. (2010) *Merchants of Doubt: How a Handful of Scientists Obscured the Truth from Tobacco Smoke to Global Warming*, New York: Bloomsbury Press.

Oreskes, N. (2022) 'Why didn't they act?' in Thunberg, G. ed. *The Climate Book*, Chapter 1.7.

Peters, G. (2022) 'The rise of renewables', in Thunberg, G. ed. *The Climate Book*, Chapter 4.6.

Piketty, T. (2020) *Capital and Ideology*, Bellknap Press of Harvard University, Cambridge.

Ritchie, P.D.L., Clarke, J.J., Cox, P.M. and Huntingford, C. (2021) 'Overshooting tipping point thresholds in a changing climate', *Nature* 592: 517–523.

Semmens, A. (2001) 'Maximising justice for environmental refugees', in Gleeson, B. and Low, N. eds. *Governing for the Environment, Global Problems, Ethics and Democracy*, Basingstoke: Palgrave, Chapter 5, pp. 72–87.

Shepherd, D. (2021) 'Ten worst disasters cost US$170 billion dollars', *The Age*, Melbourne 28/12/2021.

Shindell, D. (2022) 'Air pollution', in Thunberg, G. ed. Chapter 3.4 p. 140.

Stern, N. (2022) 'Emissions and growth', in Thunberg, G. ed. Chapter 4.24.

Thunberg, G. ed. (2022) *The Climate Book*, London and New York: Allen Lane/Penguin/ Random House.

Willcock, S., Cooper, G.S., Addy, J. and Dearing, J.A. (2023) 'Earlier collapse of anthropocene ecosystems driven by multiple faster and noisier drivers', *Nature Sustainability*. https://doi.org/10.1038/s41893-023-01157-x (accessed 26/06/2023).

8

DEMOCRACY AND THE INTERNATIONAL ORDER

Introduction

People fight for democracy for a variety of reasons: for representation in the deliberations of the State, for freedom of expression and assembly, for protection against arbitrary arrest and punishment, for equality under the law, and for equal access to the necessities of life: education, literacy, health care, and shelter. The struggle for democracy is an inextricable part of the ongoing struggle for equality. As Piketty writes,

> Like the quest for ideal democracy, which is nothing other than the march toward political equality, the march toward equality in all its forms (social, economic, educational, cultural, political), is an ongoing process that will never be completed.[1]

The people's struggle for democracy is happening today, with terrible costs in human life in Ukraine, Iran, and Myanmar. The institutions of democracy have been upheld in the USA, Brazil, the Philippines, Australia, and Sri Lanka. So-called populists have been dismissed: Trump, Bolsonaro, Duterte, Morrison, and the Rajapaksas (Mahinda and Gotabaya). Democracy everywhere remains imperfect. It is important to distinguish between two kinds of imperfection. The first is the imperfection of the process of representation and inclusion, and the second is the imperfection of outcome: how democracy succeeds or fails in the quest for equality 'in all its forms'.

So, there are two questions: Is democracy in crisis? And is democracy serving the interests of social equality? Behind these questions are major issues of world peace and international governance for climate justice.

DOI: 10.4324/9781003382133-10

Is Democracy in Crisis?

A Swedish Institute, VDem Gothenberg, studying democracies worldwide reports an alarming decline in liberal democracies and a corresponding rise in autocracy. Liberal democracies peaked in 2012 with 42 nations, and in 2021 were down to 34. The decline, VDem reports, 'is especially evident in Asia Pacific, Eastern Europe and Central Asia, as well as in parts of Latin America and the Caribbean'.[2] However, 'among the top ten democratizing countries six transitioned from autocracy to democracy in the period'.[3]

VDem observes a trend towards 'electoral autocracies'. This definition includes countries where regular elections are held, but freedom of speech and assembly for minority groups are compromised, the independence of electoral management bodies and judiciary is undermined, and repression of civil society is increased. These 'anti-pluralist' countries (in the VDem analysis) include Brazil, Hungary, India, Poland, Serbia, and Turkey. Nevertheless, since 2021 Brazil turned back to democracy in 2023, and potentially Thailand. There has been an upsurge of united opposition in Turkey (though President Erdogan won the 2023 election) and a massive backlash against authoritarian tendencies in Israel.

Though the VDem analysis includes a complex range of indices, a precise definition eludes the theory of 'democracy'. So, it is perhaps fairer to conclude that democracy is in flux. In a perceptive analysis, the German political scientist Jan-Werner Müller states, 'Everybody thinks they know that democracy is in crisis, but how many of us know what democracy is?'[4] So, why do we think democracy is in crisis, and what exactly are the defining features of democracy? These are the questions Müller goes on to answer.

Fake Democracy

The most obvious evidence of crisis occurred in January 2021 when, following the 2020 presidential election in what is held to be the very seat of democracy, former president Donald Trump encouraged thousands of his followers to march on the Congress and fight to restore Trump, the loser, to power. The Trump mob invaded the Capitol, swarmed over police defenders, and allegedly tried to find Democrat leaders to kill.

But there are other signs of crisis in what Müller calls 'fake democracies', in countries headed by authoritarian-populist leaders including Viktor Orbán in Hungary, Recep Tayyip Erdogan in Türkiye (formerly Turkey), Jarosław Kaczyński in Poland, Narendra Modi in India, and Jair Bolsonaro in Brazil.[5] Donald Trump, of course, and (perhaps) Benjamin Netanyahu in Israel are also caught in the authoritarian-populist net.

The focus of Müller's critique is the 'authoritarian-populist art of governance' whereby these right-wing leaders seek first to win elections, then alter political institutions to maintain themselves in power. While acknowledging the considerable differences amongst the countries listed, broadly speaking, Müller argues, the authoritarian-populist art of governance is based on nationalism (often with racist overtones), on hijacking the state for partisan loyalists, and, less obviously, on weaponizing the economy to secure political power: a combination of culture war, patronage and mass clientelism.[6]

Interestingly, Müller discounts the possibility of authoritarian-populist governance as a precursor of fascism. The latter he defines strictly as the mass mobilisation and militarisation of entire societies. Today, he argues, 'while hatred against vulnerable minorities is being fanned, no systematic cult of violence that glorifies mortal combat is being instituted'.[7] Such a strict definition does not allow for the very broad examples of fascism, or at least fascist precursors, identified by Polanyi (see Chapter 2).

Moreover, Müller later acknowledges that in Nazi Germany there were many areas of life that proceeded as normal, with marriages concluded and business contracts written and enforced. Yet the normal life was accompanied by the 'prerogative state'.[8] Vladimir Putin's Russia fits Ernst Fraenkel's concept of the Fascist 'dual state' almost perfectly: 'split between ordinary rule-governed life and unpredictable repression'.[9] Not only overt opponents of the war on Ukraine are imprisoned or murdered, 23 extremely wealthy men from the business, military, and media spheres died under mysterious circumstances in a single year, 2022, the year when Russia invaded Ukraine.[10]

If authoritarian populists (fake democrats) come to power with the 'will of the people', in what ways are they not democratic? Müller proposes that there is a 'hard border of democratic conflict' distinguishing 'fake democracies' from 'real democracies': a democratic people cannot expel or disenfranchise citizens (let alone murder them), excluding them from governing institutions against their will. A democratic people cannot deny the standing of particular citizens 'for that conflicts with a commitment to democratic equality'.[11] If that principle is not accepted by those who govern, however they are elected, they stand outside the hard border that defines democracy.

Real Democracy

Müller then embarks on a discussion of the constituents and infrastructure of 'real democracy' looking back at the historical origins of democratic theory from Athens to the liberal philosophers. Some of the points he makes are familiar. All public policy is contestable. Democratic outcomes are essentially uncertain. Respect for losers must be maintained. But there has to be trust in the core political institutions. Losers (like Trump and Bolsonaro), who claim that they lost an election because those institutions are corrupted, fatally undermine that trust. There has to be some tolerance of civil disobedience that does not undermine trust. Political parties and communication media form the critical infrastructure of democracy.

All of this is an argument for the kind of democracy that Americans thought they had, a *pluralist* democracy in which policy outcomes are determined by the political struggle amongst many and diverse competing interests.[12] A system in which there is always room for collaboration and compromise amongst interest and pressure groups. Franklin Delano Roosevelt's New Deal would have been impossible without collaboration. So would the American contribution to the Allied victory in World War II (as noted in Chapter 3).

In recent years, that has changed. Müller remarks perceptively that the rise of populist leaders is not explained by people craving the 'strong hand' of authoritarian leadership: 'the problem is not that ordinary people wish to be done with democracy; the problem is the choices on offer in highly polarised and increasingly fragmented societies' – a *poverty* rather than a plurality of choices.[13] Müller points to a 'double secession'. Secession, that is, from the pluralist politics that makes possible collaboration amongst interests.

The first secession Müller explains is 'to put it bluntly, that of the most privileged'.[14] These are not 'the liberal, cosmopolitan elites', demonised by Trump as located in the heart of big government and the core of the Democratic Party. They are rather the hyper-rich who tend, if anything, to be Trump supporters. They are cheerleaders of globalisation; they 'appear to be able to retreat from any real dependence on the rest of society'[15] ('appear' because they still rely on workers and the state to maintain their independence and wealth). They live in wealthy enclaves, if not gated communities, and 'reproduce many of their privileges over the generations'.[16] They use 'concierge medical services to have access to doctors, nurses, and of course testing with results available within thirty minutes, all reserved exclusively for them'.[17]

Most importantly, they conceal their wealth in 'more or less glamorous tax havens' and use 'transnational accounting tricks with entirely stateless entities'. Citing the two distinguished economists Saez and Zucman:

> US firms have in 2016 … booked more than 20 per cent of their non-US profits in 'stateless entities' – shell companies that are incorporated nowhere, and nowhere taxed. In effect they have found a way to make $100 billion in profits on what is essentially another planet.[18]

Müller questions whether the USA is today a *de facto* oligarchy. The hyper-wealthy have been able to secede from any responsibility for the society in which they live. Yet they can 'deploy their wealth to create dependencies and distort the political process'.[19] In the USA especially (but by no means only in the USA), funding for political parties is heavily dependent on wealthy donors, who can of course occasionally enter politics themselves.

The 'second secession' refers to the increasing number of citizens at the lower end of the income spectrum who no longer vote. They see no escape from their circumstances: 'for them – there is only the land of deprivation and quiet despair, which can be as much in people's minds as on maps of what are sometimes euphemistically called "disadvantaged communities"'.[20]

Müller quotes the German political scientist Claus Offe to describe the political logic at work:

> As people are conditioned to 'waste' their rights and political resources, and as competing political elites and political parties come to understand that parts of the electorate are less likely to make use of their political resources, those elites will concentrate their platforms, campaigns and mobilization strategies upon those segments of the citizenry who actually 'count' and neglect others.[21]

This 'invisible' secession can become reinforcing. Political parties have no reason to shape policies that attend to the needs of those who do not vote. This strengthens the view of the poor that there is nothing for them in current politics. Thus, political platforms and voter mobilisation strategies are shaped to meet the demands of those citizens 'who actually count'. Müller concludes that the diverging preferences of the 'well-to-do' and the rest of the population mean that in the USA, on many issues, the political system is unresponsive to the entire bottom two-thirds of the citizenry.

Müller's analysis is focused on the USA, but the concept of a double secession can also be applied to other democracies, notably Britain and France. Though the yellow vest riots of 2018 and the anti-racist riots of 2023 (in France) had quite different triggers, the sense of political powerlessness and 'secession' seems to be common to both.

Müller, however, warns against oversimplification, acknowledging that a different kind of poverty – poverty of opportunity – touches the increasingly stressed professional middle classes as their prospects for advancement via education and higher paying jobs slip away. The collapse of the belief in intergenerational progress, according to the Polish sociologist Adam Przeworski, is 'a phenomenon at a civilizational scale',[22] which helps explain the broad appeal of 'fake' democrats such as Trump and Orbán.

Democracy's Historical Purpose

We have to be careful, as Müller's commentary indicates, not to neglect the social purpose of democracy – the march towards equality. Democracy can certainly be regarded, politically, as an end in itself, providing for inclusion and social cohesion, and a sense of political fairness and stability, allowing many voices to be heard and become effective. But, as we saw in Polanyi's historical analysis, democracy is a means to other social ends: social justice, better working conditions, and the defeat, in William Beveridge's terms, of the five giants: Want, Disease, Ignorance, Squalor, and Idleness.

Up until the neoliberal regression took hold (even before the arrival of the Thatcher Government), the belief that democracy could and would deliver those social ends was strong. It was a belief that broke the ideological power of class polarisation. It was a unifying counterforce which gave rise to a comforting belief in pluralism as the basis of liberal democracy. The erosion of that belief has resulted in new forms of polarisation: ethnic, nativist, and above all economic. However, to label democracies 'fake' that do not at present live up to the criteria of 'real' democracy is a false polarisation.

Müller's analysis of democracy is focused on basic principles and on political leaders who appear to breach those principles. In the historical perspective, it is wrong to label democracies 'fake' on the basis of the behaviour of their current political leaders. Therefore, we need to examine not only the historical context but also the constitutional context in which political leaders operate. To illustrate the point, we will look briefly at three cases: Britain (with the oldest Parliament), the USA (the self-described 'leader of the free world'), and India (the world's largest democracy). These three are arguably 'real' but *flawed* democracies. Here, we set aside, for the moment, the issue of inequality, poverty, and 'secession'.

Britain

In Britain, the history of democracy is often believed to begin in 1215 with Magna Carta following a rebellion against a wayward and unpopular king (John) on the part of a group of military aristocrats (the barons). But modern democracy was only really sealed with the disempowerment of the House of Lords in 1909 (discussed in Chapter 3). Until then, economic liberalism was established under the tutelage of aristocracy, not in the modern sense of democracy.

Even today, the convention that the winner of a seat in the House of Commons, determined by whoever wins the largest proportion of votes in a constituency,[23] delivers unbalanced power to the two dominant political parties. In 2019, the Green Party won 2.7 per cent of the popular vote nationwide and just a single seat. If parliamentary representation were commensurate with votes cast nationwide, the Green Party would have won 17 seats.

In 2019, the Conservative Party won 365 seats with 42.4 per cent of the vote. If seats were commensurate with votes cast nationally, the Party would have won 276 seats while the Labour Party would have won 260 seats (instead of 262). The Liberal Democrats would have won 75 seats (as opposed to just 11) and the Scottish National Party 25 seats. Therefore, a Coalition of the broad Left (Labour, Lib-Dems, Scottish Nationals, and Greens) would have won 377 seats, an overall majority of 58 per cent to 42.4 per cent and a margin of 101 seats over the Conservative Party.

Even Margaret Thatcher's so-called 'overwhelming' victory over Labour in the 1979 election, if commensurate with votes cast nationally, would have delivered a working majority of 24 seats to a Labour–Liberal Coalition (with a lead of 51.6%, to the Coalition, over 47.2% to the Conservative Party).

The United States of America

Democracy in the USA, which also operates under the first-past-the-post voting system, is undermined by the practice of State legislatures manipulating the boundaries of electoral constituencies to favour the party in power (gerrymandering) and by placing restrictions on voting. A report from the Brennan Center for Justice noted that in 2021 nineteen states have enacted draconian laws that will make it harder to vote. Some Republican-dominated States impose criminal penalties on election officials and other individuals for engaging in ordinary, lawful, and often essential assistance for voters. Such tasks include handing out water to people standing in line for long periods in hot weather, helping voters who have disabilities, and regulating the conduct of poll-watchers. Even though 25 states have made it easier to cast ballots, this does not restore the balance nationally.[24]

The Constitution of the United States establishes that each State should be equally represented in the Senate, with two Senators, regardless of its population size. Article 5 of the Constitution (on Amending the Constitution) prohibits any State from being deprived of equal representation in the Senate without its (the State's) permission. Because the population size of States varies greatly, the principle of 'one person, one vote' has never passed the Senate.

The Constitution was designed in the aftermath of the Revolution against Britain. Despite the rhetorical flourish of the opening words of the document ('We, the People of the United States'), the founding fathers of the Constitution did not want a return to revolutionary ideas. Rather, in the summary drafted by the aristocratic Gouverneur Morris, the goals of the Constitution were to be: 'a more perfect union, justice, domestic tranquillity, common defence, general welfare, and personal liberty and prosperity'. Commenting on deficiencies of the Constitution, BBC Washington correspondent Nick Bryant notes that the whole point of the Constitution was to guard against the tyranny of the majority and an excess of democracy.[25]

The right to vote was not originally written into the Constitution as a positive right. Voting was at first expected to enfranchise only white men of property. After the Civil War, voting rights were extended in a negative way (voting shall not be denied) to men regardless of race, colour, or previous condition of servitude (the 15th Amendment). In 1920, after a long campaign for women's suffrage, the 19th Amendment was passed stating that 'the right of citizens to vote shall not be denied or abridged by the United States or by any State on account of sex'.

There is the questionable separation of powers in the USA following the reversal of the Roe versus Wade judgement by the Supreme Court and other subsequent judgements that clearly usurp the will of the people. Alexander Hamilton in *Federalist Paper 78* (1788) wrote of the separation of powers:

It equally proves, that though individual oppression may now and then proceed from the courts of justice, the general liberty of the people can never be endangered from that quarter; I mean *so long as the judiciary remains truly distinct from both the legislature and the Executive.*

Article 3 of the Constitution provides that the President of the USA nominates a justice and the Senate provides 'advice and consent' before the person is formally appointed to the Court. In a highly polarised polity in which the President and a Senate majority belong to the same Party, this erodes the principle of the separation of powers between the executive and the judiciary at the highest level.

The Republican Party has lost the nationwide vote in seven of the last eight elections yet has retained the power to block legislation that its supporters do not like.[26] Arguably the USA has turned from being a pluralist into a *minoritarian* democracy. This is particularly clear in the matter of gun violence.

1.5 million civilian Americans died from a gunshot between 1968 and 2017, more than the number of soldiers killed in every war since the American War of Independence in 1775. In one year, 2020, 45,000 Americans died from gun violence (more than 19,000 of which were homicides) – an increase of 43 per cent over the figure for 2010.[27] For every 100 residents, there were on average 120 firearms owned. Compared with other countries, the USA is quite exceptional in firearm homicides with 4.12 homicides per 100,000 people. The next highest rate was in Chile with 1.82. The average for the next 19 countries was 0.29. The UK and Australia rated 0.04 and 0.18, respectively.[28]

The failure of the USA to prevent people, frequently schoolchildren, from being murdered with firearms has to be regarded as an egregious failure of democracy. *No citizen* can want that outcome. The failure to regulate firearm ownership is a result of a combination of factors, the least of which is the Second Amendment intended to equip the new nation with a 'well-regulated militia' for defence against the British.[29] But the deployment of funding to both political parties by the gun industry, together with the blocking strategy of the Republican Party, has led to a path-dependent situation in which the more people have guns, the more they feel the need to defend themselves by owning one. Though it is obvious that owning a gun does not prevent anyone from being murdered with one.

India

Müller includes current Prime Minister Narendra Modi in his list of fake democrats. Others are even more critical. D.R. Chowdhury writing in the *New York Times* accuses Modi and his Party (Bharatiya Janata Party) of profaning Indian democracy,

espousing an intolerant Hindu supremacist majoritarianism over the ideals of secularism, pluralism, religious tolerance and equal citizenship upon which the country was founded after gaining independence on August 15th, 1947.[30]

The British international policy institute, Chatham House, is also critical of the direction in which Modi is taking Indian democracy.[31] The report outlines 'challenges to democracy' in India, many existing before the government of Narendra Modi, including corruption and criminality amongst politicians, excessive use of violence by police against demonstrators, and a dysfunctional legal system 'which leaves many languishing in detention before trial for "crimes" including peaceful protest'.[32]

Yet the greatest challenge facing India's democracy is: 'that it has failed to deliver the kind of sustained economic development enjoyed by neighbours like China over the last four decades. It has also failed to eliminate extreme poverty'. Inequality is rampant with educated elites in globalised cities like Delhi and Mumbai living 'completely different lives from India's poorest citizens'.[33]

Even before independence, nationalists had argued that India should be the homeland of Hindus, separated from Pakistan as the homeland for Muslims. Modi and the BJP represent this nationalist ideology, *Hindutva*, which has remained consistent for over a century. In 2019 Modi's government passed the Citizenship Amendment Act, which eased citizenship requirements for various religions – but expressly omitted Muslims.

The autonomous status of Kashmir, the only Muslim majority state, was revoked and thousands were arrested including politicians, activists, and separatists. In Assam, where Muslims represent about a third of the population, detention camps were set up for migrants who could not prove their Indian citizenship. '1.9 million Muslims had already been effectively stripped of their citizenship in Assam after being left off India's National Register of Citizens'.[34]

The Chatham House report concludes that

> Through its control of the media, monopolization of campaign finance and harassment of opponents, India seems set on a path to becoming an illiberal pseudo-democracy similar to Turkey or Russia. However, when the BJP has faced a united opposition in recent state elections it has generally lost.

As in Britain, India's voting system is first-past-the-post.[35]

Modi has taken India down the neoliberal path. In order to generate 'growth' and 'jobs', the BJP government has reduced taxes on the big corporates while increasing the tax burden on citizens through rises in direct and indirect taxation. The government has privatised public sector enterprises. Air India was sold to the Tata Group with the airline's debts left for the government to settle. To connect India more decisively with the global economy, Modi reduced customs duties to encourage imports, a move that has been damaging to domestic medium and small-scale industries. Income taxes rose by 117 per cent while corporation tax rose by only 28 per cent.

Journalist Subodh Varma commented in 2012,

> No number of concessions to the rich will boost the economy because what is needed is buying power in the hands of the people, not in the hands of the rich. This can be done only by increasing government spending and raising taxes on the rich. But the Modi government continues blindly in the opposite direction – causing immense and continued suffering to the people of India.[36]

As before, we need to look beyond the political leader and contemporary government at both the constitutional context and the history of India before passing judgement on the country's democracy. The democratic Constitution of India was passed by the Constituent Assembly in 1949 and came into force in January 1950.

The preamble to the Constitution states:

> We the people of India, having solemnly resolved to constitute India into a sovereign, socialist, secular, democratic republic, and to secure for all its citizens justice, social economic and political; liberty of thought, expression, belief, faith and worship; equality of status and opportunity, and to promote among them all fraternity assuring the dignity of the individual and the unity and integrity of the nation, in our Constituent Assembly this twenty sixth day of November 1949, do hereby adopt, enact and give to ourselves this constitution.[37]

India's Constitution is a federation with national and state parliaments modelled on the Westminster system. At the national level, the President of India is the formal head of the state. There is a separation of powers between the executive, headed by the Prime Minister, the legislature (parliament), and the judiciary with the Supreme Court as the highest judicial authority. There is an upper and lower house of parliament (the Rajya Sabha and Lok Sabha). There are 29 states with about 20 different official languages. Each state mirrors the parliamentary structure of the national government with an upper and lower house.

The unification of India, as a modern democratic state, is a very recent development. As Piketty points out, we need to ask how a population as large as India's with a population of 200 million by the end of the eighteenth century managed even then to co-exist peacefully as a human and political community? In his chapter on the complex and evolving social and ideological structures of the sub-continent, Piketty offers a finely detailed analysis of India's long social and ideological history going back over 2000 years.

The modern Indian Constitution was shaped at a time when 'socialism' was a dominant ideology in the colonial power, Britain. Piketty reminds us that 'It was left to independent India to achieve administrative and political unification after 1947 under a vibrant, pluralistic parliamentary democracy' – and on an unprecedented human and geographic scale. By comparison,

Europe is currently attempting to build a political organization on a large scale with the European Union and European Parliament (although Europe's population is less than half of India's and its political and fiscal integration is much less advanced).[38]

All of these three nations are instances of *flawed* rather than fake democracies. Political leaders sometimes try to reshape democratic institutions to serve their own ends. This has certainly happened in the USA with Trump and in India with Modi, arguably also with some leaders of post-communist European nations, and most recently in Israel with Netanyahu. Yet, the institutions of democracy have not succumbed. The struggle for democracy in authoritarian states continues. When democratic institutions are threatened, the public responds, if not in parliaments then in the streets. Populist leaders can quickly become unpopular.

The most important question, however, is why populist leaders attract public support to enable them to win power? The answer lies in the dual secession noted by Müller. Large sections of the public have lost hope that governments can act to advance their interests. So, they turn to leaders who offer them new slogans and explanations for their plight, or, as in France, they take to the streets.

But it is not the failure of political leaders that is the problem; it is the failure, or rather the weakness, of the state. As Daniele Archibugi argues, the existence of a recognised institution that is the only one authorised to use force legitimately is the precondition for the birth of democracy.[39] The capacity of the state has been radically weakened by the privatisation of key public monopolies (water, power, and transport), cutting the capacity of the public service and outsourcing advice to government, and deregulation of capital through international agreements. This is a theme which we will take further in the next chapter.

Economic consultancy to governments is dominated by four giant international firms: Deloitte, KPMG, Ernst and Young, and PwC (Price Waterhouse Coopers). They are all private partnerships with few reporting obligations. Across the world, governments rely on consultants. In 2021, the global consulting services market was valued between US$700 billion and $US900 billion.[40]

The potential for conflict of interest is extreme. This was realised in 2023 when it was revealed that PwC, after giving advice to the Australian Government on how to 'crack down on multinationals avoiding corporate tax through cross-border income and asset transfers', was found to sell their advice to the very targets of the crackdown, 'teaching them how to sidestep the same anti-avoidance measures PwC was helping to draft'.[41] This gross misconduct has so far (May 2023) had no legal consequences. Senior Counsel Geoffrey Watson rightly levels his criticism at the Government. 'Why on earth was PwC – a substantial contributor to the global problem of cross-border tax minimisation – involved in designing Australia's response to that very problem?'[42]

The inherent conflict of interest is not limited to economic policy. In the USA, the global consultancy McKinsey, from 2011 to 2019, served as an advisor to the Federal Food and Drug Administration (FDA) (the regulator of medicines). During that time, it failed to disclose to the FDA that it also served its client Purdue Pharma, the corporation marketing the opioid OxyContin linked to the American overdose crisis.[43] McKinsey pleaded guilty to criminal charges and agreed to a settlement of US$641 million. Criminal and civil investigations into Purdue and its shareholders resulted in the dissolution of the company, with a civil settlement of US$8 billion.[44]

These scandals in Australia and the USA have had repercussions around the world. Suddenly, the public has become aware of the power of these consulting firms. The policies that they, and the myriad of smaller agencies, recommend to governments will not be detrimental to (and may even advance) the interests of their private clients, nor will their advice depart from neoliberal orthodoxy. In effect, the key parts of the public service in some Western democracies have been captured by the interests of big capital and neoliberal ideology.

Meanwhile, there are burning issues of international governance. The most immediate is whether a new global conflict can be avoided between Russia, aligned with China and possibly North Korea, and the democracies: The European Union, the USA, Canada, UK, Japan, Australia, New Zealand, and South Korea. The second is the equitable long-term international distribution of the costs of climate change.

The End of the Peace Interest?

Karl Polanyi noted that a hundred years of peace in Europe prevailed from the end of the Napoleonic wars until the beginning of World War I. World trade required peace, and the 'Great Powers' strove to maintain it. Recall (from Chapter 2) that world peace depended on a stable monetary system based on the gold standard, holding together international trade.

At the end of 2022, the Nobel Prize winning economist Paul Krugman, writing in the New York Times, posed the question, 'Is this the end of peace through trade?'[45] The British journalist Norman Angell published a book in 1909 that argued that the cost of war was so great that no nation could possibly gain by starting one. Krugman points out how wrong Angell was, just five years before the start of World War I. With the wisdom of hindsight, Polanyi described the circumstances of the global economy that led to the end of peace.

Krugman reviews the contemporary circumstances in the world economy that pose a threat to 'peace through trade', based on an agreement among trading nations that has prevailed since the end of World War II. Under the aegis of the United Nations, the General Agreement on Tariffs and Trade (from 1948) and its successor the World Trade Organization (from 1994) have successfully promoted international trade by reducing or eliminating tariffs and other trade barriers.

Now the USA, 'which largely created the world trading system', has started imposing restrictions in the interest of national security – first under Trump and continuing under the Biden presidency. The former Australian Prime Minister and ambassador designate to the USA, Kevin Rudd, commented on Bloomberg TV that the political class in America 'was now dominated by an overriding protectionist sentiment' and urged Congress to do more to open American markets to allies in Asia and Europe.[46]

The USA, Britain, Canada, and the European Union have enacted economic sanctions on the People's Republic of China in response to human rights violations by China against Uighur people in Xinjiang Province. China retaliated in kind.

Krugman asks whether 'peace through trade' or 'the peace interest' (Polanyi) really applies to authoritarian governments. Krugman cites the case of Taiwan: 'Are we sure that the deep integration of Taiwan into China's manufacturing system rules out any possibility of invasion?'[47] Russia's war on Ukraine has triggered aggressive trade sanctions on both sides (Western democracies and Russia and its allies), completely disrupting the world economy. Restrictions on trade with Russia are the principal weapons of EU countries to punish Russia for its invasion. Russia responded by cutting gas and oil supplies.

Krugman concludes that the current world order

largely reflects strategic considerations: leaders, especially in the US, believed that more or less free trade would make the world safer, and more amenable to their countries' political values. But now even relatively internationalist policy makers, like officials in the Biden administration, aren't sure about that. This is a very big change.[48]

The possible end of the peace interest may seem to have little to do with the climate crisis, until we think about the question of international cooperation. Both China and Russia are critical players in the effort to mitigate global heating. The Ukraine war has brought the two countries closer together. It is possible that an authoritarian axis is beginning to form, including also Iran, North Korea, and possibly Myanmar. The relevance of the United Nations Security Council is questionable now that one of its members, Russia, is engaged in an illegal war with a sovereign democratic nation and another, China, is continually threatening Taiwan (which it claims as its province). Economic competition is already threatening to merge into a 'cold war' between the USA (and its allies) and China in the Pacific.

In Chapter 5 we saw how 'shock therapy' devised at the height of the neoliberal regression by international advisors from the West in effect exported capitalist oligarchy to Russia and the entire communist bloc. The war on Ukraine has once again polarised the world system: this time between 'autocracy and democracy'.

The Russian political geographer, Igor Okunev (mentioned in Chapter 3), argues that the annexation of the Crimea by Russia in February 2014 marked the beginning of a systemic crisis of the international order.

> This system was based on the right and commitments of the World War II victors to work out an international order and act as its guarantors. However, the role of great powers is not only in sustaining the world but also in perfecting it. They are responsible for changing the world system.[49]

It is becoming clearer by the day (2023) that NATO is at proxy-war (rather than 'cold war') with Russia in Ukraine. Putin's regime started the war, driven by the humiliation of Russia by the West under the influence of the neoliberal regression. The West cannot now deny the necessary weapons to Ukraine to win the war and, as of the time of writing, appears determined to supply weapons indefinitely. However, a wider perspective on the degradation of the international order raises questions both about the future of the current state of democracy and of international governance.

The Climate Crisis and the International Order

The United Nations was formed after World War II to address issues of transnational importance. An extensive network of treaties, conventions, and legal principles has been created under the aegis of the UN to regulate environmental exploitation. This 'negotiated order' and its efficacy have been discussed and debated at length.[50] But perhaps the best-known institution of the United Nations is the Security Council, created with the agreement of the victorious powers after World War II to preserve world peace.

The functions of the Council are specifically directed at abating military threats to peace and negotiating alternatives to armed conflict amongst nations. The Council is even empowered to take military action against an aggressor. The continual presence of international armed conflicts since its creation (from the Korean War to the wars in Vietnam and Iraq, and now Ukraine) testifies to the very limited success of the Security Council in preserving peace. But the Council has up to now at least presided over a long peace between the great powers. Today, with increasing belligerence from Russia and China (permanent members of the Security Council), matched by equal belligerence from the USA and the NATO alliance, even that success is now in question.

It is certainly too soon to predict the end of the interest of the great powers in preventing unlimited war. The consequences of such a war for the economic interests of China, the USA, and Europe are probably too enormous, even without the horror of a nuclear exchange. However, today, and increasingly in the future, the developing climate crisis puts a different kind of security at stake: environmental security. In an age when homes, businesses, and livelihoods throughout the world are threatened by changing weather driven by global heating, how might the threat of environmental insecurity be addressed at the international level?

In the 1990s, the broad question of environmental justice was widely discussed, bringing together a range of issues of human–environment relations at local, national, and international levels. We were warned 20 years ago that the economic crises of the 1970s and 1980s were resulting in economic costs being displaced into the 'realm of nature' in the 1990s and 2000s: 'the short-term economic health of the salariat and corporate owners is being increasingly secured through the long-term sacrifice of the environmental health of the subaltern – peoples of colour and the poor (including developing world peoples)'.[51]

The processes of restructuring facilitated by a slew of international agreements to facilitate global movement of capital have enabled businesses to extract more value from human labour and the natural environment at reduced time and cost. Global heating, as we now understand it, brings into even sharper focus the connection between the social and environmental crisis: the unequal impact of climate on human habitats and human social strata.

Among these discussions of environmental justice, the potential for global or 'cosmopolitan' governance was raised. While the dangers of global heating could be projected into an indefinite future, these discussions around 'cosmopolitanism' remained at a theoretical level. In the 2020s that future has begun to happen, and it is becoming clear that the basic need for global climate security is not being met by existing institutions of international governance. So, we need to reconsider where the concept of cosmopolitan democracy might lead.

Institutional change in favour of cosmopolitan governance needs to be preceded by a widespread sense of 'environmental citizenship' or 'cosmopolitan citizenship' with a vision of the planet as the ultimate homeland. This concept does not need to override local and national cultural conceptions of homeland, though in practice they may conflict.

The political philosopher Janna Thompson has pointed out that citizenship in Western democracies encompasses social rights: 'the right to benefits and services that enable individuals to become or remain agents capable of autonomy'.[52] The intergenerational perspective is crucial to her argument. As individuals, we care about our children and their children. As we become aware that the places in which we live are changing in dangerous ways because of weather driven by climate change, our small local world suddenly becomes large. We are forced to think of the future of our offspring and the environment in which they will live.

Planetary citizenship must encompass *their* environmental rights, the protection of which entails political action. Thompson argues: 'A planetary citizen is someone who assumes her share of responsibility for the collective achievement of goods which she and virtually everyone else values',[53] that means someone who wants to be a political actor, not just a believer. Ideology is crucial, as Thomas Piketty argues.[54] But cosmopolitan citizenship, understood in this way, is *not only* a matter of belief or ideology, it is ideology *internalised*: it is a matter of our very *being* – a matter of ontology.

What stops people from acting as citizens whose sphere of concern extends beyond the individual and immediate family, whether locally or globally? Individualism, often blamed as responsible for our social problems today, is not just a matter of ideology but of perceived necessity. It is not possible to be a cosmopolitan citizen if you have seceded from political life because the sphere of your life has shrunk to the necessity of survival of yourself and your family. It may also be difficult to be a cosmopolitan citizen if your life is entirely dominated by materialist imperatives, no matter how well off you are.

Of course, there are exceptions to this 'perceived necessity'. There are people at both ends of the economic spectrum who act as citizens: poor people who nevertheless spend time and money to help others in their community; rich people who give away significant portions of their wealth for humanitarian and environmental causes. But there are many whose social condition stands in the way of citizenship behaviour.

Democracy is ultimately majoritarian. Transformative political action requires a widely shared sense of citizenship, of social responsibility, across social strata and nationalities, and supportive of institutional change *before* institutional change can happen. The history of the transformation of liberalism into social democracy discussed in Chapters 2 and 3 suggests that social fairness – freeing actors to be citizens – precedes institutional change, then at the national level, today at both national and global levels. Yet, real social change requires institutional change accomplished by governments and their leaders.

Citizenship and the Covid Crisis

The crisis of Covid-19 has been something of a test of global citizenship and institutions. As of 21 February 2023, there had been 757,264,511 confirmed cases reported worldwide and 6,850,594 deaths.[55] Though statistical evidence of the socio-economic impact of the virus is scarce, anecdotal reports tell us that the greatest impact of the virus has fallen on the poor and people of low income.

The image of impoverished people in India struggling to get hold of oxygen cylinders for their dying kin stays in the mind. In the early period of 'lockdown' to prevent movement of people spreading the virus, those with insecure jobs, often contracted by labour hire companies, had to choose between their livelihoods and public safety. In Australia, hundreds of old people in privately operated care homes died from infection brought in by their care workers on labour hire contracts.

For countries, the case–fatality rate (percentage of deaths per number of cases) provides some indication of the capacity of each nation's health system to save lives. Of the 37 nations with a case–fatality rate above 2.0 per cent, most are poor nations. The highest rates are those of Yemen (18.1%), Sudan (7.9%), Syria

(5.5%), and Somalia (5.0%). All of these have suffered from civil (and, in the case of Yemen and Syria, international) violent conflict. Seventeen of the 2.0 per cent plus nations are in Africa. Five are in Latin America, including Peru with 4.9 per cent. Five are in East Asia: Myanmar (3.1% also afflicted by civil war), Indonesia (2.4%), Cambodia (2.2%), and China (2.1%). Three are in Caribbean nations: Haiti (2.5%), Trinidad and Tobago (2.3%), and Jamaica (2.3%). Four are European nations: Bulgaria (2.9%), North Macedonia (2.8%) %), Hungary (2.2%), and Ukraine (2.1%).[56]

Within nations, there is now a considerable body of published research that shows that the severity of impact of Covid-19 is closely related to socio-economic circumstances, class, race, and gender.

For example, a nationwide study of the effects of socio-economic position, race, and gender in the USA on Covid-19 mortality found that mortality was five times higher for low versus high socio-economic position (SEP) adults (with 72.2 deaths per 100,000 versus 14.6 deaths per 100,000 high SEP).

Further, 'The joint detriments of low SEP, Hispanic ethnicity, and male gender resulted in a COVID-19 death rate which was over 27 times higher than for white (non-Hispanic) women'.[57] A study of Covid-19 in India found that 'Illiterate and less educational (socio-economic) status patients with COVID-19 in India have significantly greater adverse in-hospital outcomes and mortality. This is related to more severe disease at presentation'.[58]

The historian Adam Tooze in his account of Covid-19 (published in 2021) points out that 'given the limitation of our social, cultural, and political coping capacities, we depend on techno-scientific fixes' to solve problems of global scale.[59] Tooze is no techno-optimist. We can marvel at the speed and success of the invention and production of vaccines yet, as he observes, there is a disproportion between the scale of the crisis and the scale of the means used to resolve it. The cost of the crisis to the world economy ran into tens of *trillions* of dollars of damage. While only tens of *billions* of dollars were spent on the vaccines and 'even less to ensure their efficient deployment and fair distribution'.

Fighting the economic crisis was left to a large extent in the hands of the central banks. As Tooze writes, 'What has made central banks into the exemplar of modern crisis fighting is the vacuum created by the evisceration of organized labour, the absence of inflationary pressure, and, more broadly, the lack of anti-systemic challenge'.[60]

Tooze's book was published before the resurrection of inflation caused by the profligacy of those very banks during the Covid crisis. Rising interest rates now seem set to generate a worldwide housing crisis for those who do not already own houses or have purchased unrealistically large mortgages on the expectation of low interest. The central banks claim that manipulating interest rates is *their* only tool to fight inflation, but it is not *the* only tool open to governments. The issue of inflation and the role of central banks are closely connected with those of unemployment, poverty, and inequality to which we turn in the next chapter.

In summary then, the international order, despite the best efforts of the United Nations, is unprepared for the crises that will undoubtedly occur as climate change progresses, with ever more devastating weather events among the longer-term changes (such as destruction of coastal settlements and desertification of entire regions) brought on by the biosphere's adaptation to increasing human-generated concentrations of atmospheric carbon.

The social transformation that has to happen to address the climate crisis requires further advances in democracy reaching from the local and national levels to institutions of global governance. Progressive social transformations, as we have argued – in the steps of Polanyi and Piketty – occur as a result of 'strong mobilizations and power relationships'.[61] Following the crisis of two world wars, the power of national states was brought under democratic control to act decisively to reduce poverty and inequality. That power was drastically weakened by the neoliberal regression to which the world is still subject even as the ideology of liberalism is crumbling.[62]

Notes

1 Piketty, T. (2022) p. 18.
2 VDem Gothenberg (2022) Executive Summary, p. 6.
3 Ibid., p. 7.
4 Müller, J-W (2022). Müller is Professor of Politics at Princeton University, USA.
5 In 2023, the list would include Vladimir Putin in Russia.
6 Müller (2022), p. 4.
7 Ibid., p. 5.
8 Ibid., p. 10–11 citing the German political scientist and lawyer Ernst Fraenkel.
9 Ibid., p. 11
10 Fellner, C. (2022) pp. 4 and 5.
11 Müller (2022), p. 39.
12 Described typically in the work of political scientists R.A. Dahl (1961) and R.J. Waste (1987) Both are discussed at length in Low, N.P. (1991) Chapter 5.
13 Müller (2022), p. 21.
14 Ibid., p. 23.
15 Ibid., p. 24.
16 Ibid., p. 24.
17 Ibid., p. 25.
18 Ibid., p. 26. Citing: Saez, E. and Zucman, G (2019) pp. 77-78.
19 Müller (2022), p. 28.
20 Ibid., p. 31.
21 Offe, C. (2013) cited by Müller (2022), p. 31.
22 Müller (2022 p. 33).
23 The so-called 'first-past-the-post' system.
24 https://www.brennancenter.org/our-work/research-reports/voting-laws-roundup-october-2021, (accessed 17/01/2023). Attempts by the Democratic Party to legislate to protect citizens' voting rights have been repeatedly blocked by Republicans in Congress.
25 Bryant, N. (2020).
26 The Republican Party lost the popular vote in the presidential elections of 2020, 2016, 2012, 2008, 2000, 1992, 1996. https://en.wikipedia.org/wiki/List_of_United_States_presidential_elections_by_popular_vote_margin (accessed 18/01/2023).
27 https://www.bbc.com/news/world-us-canada-41488081 (accessed 18/01/2023).

28 Institute for Health Metrics and Evaluation (IHME) *Gun violence in World Bank high-income countries and territories with populations of 10 million or more*, https://www.healthdata.org/acting-data/gun-violence-united-states-outlier (accessed 18/01/2023).

29 The American 'militia' has grown into the most powerful army in the world!

30 Chowdhury, D.R. (2023).

31 Price, G. (2022) (accessed 23/01/2023).

32 Ibid.

33 Ibid.

34 Ibid.

35 Introduced by the British and chosen to continue by the Indian Constituent Assembly for India's post-colonial Constitution. https://aceproject.org/main/english/es/esy_in.htm (accessed 27/05/2023).

36 Varma S. (2012). Data used by Varma is sourced from budget documents on the Finance Ministry's website.

37 https://indiankanoon.org/doc/237570/ (accessed 23/01/2023).

38 Piketty, T. (2020) pp. 306–307.

39 Archibugi, D. (2001) p. 201.

40 Tadros, E. and Wootton, H. (2023) *Big Four Consulting Firms Data Tracker*, Australian Financial Review, https://www.afr.com/companies/professional-services/big-four-consulting-firm-data-tracker-20200207-p53yp4 (accessed 17/05/ 2023)

41 Watson, G. (2023) p. 24. Geoffrey Watson is a Senior Counsel and director of the Centre for Public Integrity (Australia).

42 Ibid.

43 Nichols, S. and MacDonald, K. (2023) 'Governments are increasingly reliant on consulting firms. Critics says it's often to their detriment', ABC News, https://www.abc.net.au/news/2023-03-16/australia-reliance-consulting-firms-high-cost-problem-government/102091810/ (accessed 05/03/2023).

44 https://www.justice.gov/opa/pr/justice-department-announces-global-resolution-criminal-and-civil-investigations-opioid (accessed 25/05/2023). The settlement resolves allegations from 2010 to 2018, and Purdue caused false claims to be submitted to federal health care programmes.

45 Krugman, P. (2022) p. 33.

46 Knott, M. (2023) 'US allies "thrown under a bus": Rudd', Melbourne: The Age 05/01/2023 p.8.

47 Krugman, P. (2022) Ibid.

48 Ibid.

49 Igor Okunev has a Masters' degree from the University of Manchester and a doctorate from the respected Moscow State Institute of International Relations. (https://www.e-ir.info/2021/05/25/interview-igor-okunev/). The citation is from Okunev, I (2014).

50 The negotiated environmental order is discussed in some detail in Low, N. and Gleeson, B. (1998) pp 178–183.

51 Faber, D.R. and McCarthy D. (2003) pp 38–63.

52 Thompson, J. (2001). Thompson expanded this discussion in her book *Intergenerational Justice: Rights and Responsibilities in an Intergenerational Polity* (Routledge, 2009).

53 Thompson, J. (2001) p. 145.

54 Piketty, T. (2020).

55 World Health Organization, https://covid19.who.int/ (accessed 24/02/2023).

56 Ibid.

57 Pathak, E.B., Menard, J.M. et al. (2022).

58 Sharma, A.K., Gupta, R. et al. (2021).

59 Tooze, A. (2021) p. 292.

60 Ibid., p. 293.

61 Piketty, T. (2022) p. 227.

62 As this book goes to print a new war in the Middle East is occurring following the murderous attack by Hamas on Israeli citizens and the bombing of Gaza by Israel.

References

Archibugi, D. (2001) 'The politics of cosmopolitan democracy', Chapter 13 in Gleeson, B. and Low, N. eds. *Governing for the Environment, Global Problems, Ethics and Democracy*, Basingstoke: Palgrave, pp. 196–210.

Bryant, N. (2020) *When America Stopped Being Great, A History of the Present*, London and New York: Viking Books.

Chowdhury, D.R. (2023) 'Modi's India is where global democracy dies', *New York Times*, Guest Essay, https://www.nytimes.com/2022/08/24/opinion/india-modi-democracy .html (accessed 23/01/2023).

Dahl, R.A. (1961) *Who Governs? Democracy and Power in an American City*, New Haven: Yale University Press.

Faber, D.R. and McCarthy D. (2003) 'Neo-liberalism, globalization and the struggle for ecological democracy: Linking sustainability and environmental justice' (p. 39), Chapter 2 in Agyeman, J., Bullard, B. and Evans, B. eds. *Just Sustainabilities, Development in an Unequal World*, pp 38–63.

Fellner, C. (2022) 'How the deaths of 23 Russians sparked a global mystery', *The Age*, Melbourne 03/01/2023 pp. 4 and 5.

Knott, M. (2023) *Us Allies "Thrown Under a Bus": Rudd*, *The Age*, Melbourne 05/01/2023 p. 8.

Krugman, P. (2022) 'Is this the end of peace through trade'. *New York Times* 13/12/2022, reprinted in *The Age*, Melbourne 28/12/2022 p. 33.

Low, N.P. (1991) *Planning, Politics and the State, Political Foundations of Planning Thought*, Boston and London: Unwin Hyman.

Low, N. and Gleeson, B. (1998) *Justice, Society and Nature, An Exploration of Political Ecology*, London and New York: Routledge.

Müller, J.-W. (2022) *Democracy Rules*, Dublin: Penguin Books.

Offe, C. (2013) 'Participatory inequality in the austerity state: A supply side approach', in Schäfer, A. and Streeck, W. eds. *Politics in the Age of Austerity*, Cambridge: Polity Press.

Okunev, I (2014) 'Different realities, the crimean crisis exposing the decline of the world order', *Russia in Global Affairs* 12(2).

Pathak, E.B., Menard, J.M. et al. (2022) 'Joint effects of socioeconomic position, raced/ ethnicity and gender on Covid 19 mortality among working age adults in the United States', *International Journal of Environmental Research and Public Health* 19/9: 5479, https://doi.org/10.3390/ijerph19095479 (accessed 01/03/2023).

Piketty, T. (2020) *Capital and Ideology*, Cambridge: Bellknap Press of Harvard University.

Piketty, T. (2022) *A Brief History of Equality*, Cambridge: Bellknap Press of Harvard University.

Price, G. (2022) 'Democracy in India', Chatham House Independent Policy Institute, https://www.chathamhouse.org/2022/04/democracy-india (accessed 23/01/2023).

Saez, E. and Zucman, G. (2019) *The Triumph of Injustice: How the Rich Dodge Taxes and How to Make Them Pay*, New York: Norton.

Sharma, A.K., Gupta, R. et al. (2021) 'Socioeconomic status and Covid 19 related outcomes in India: Hospital based study', *British Medical Journal*. https://doi.org/10.1101/2021.05 .17.21257364 (accessed 01/03/2023).

Thompson, J. (2001) 'Planetary citizenship: The definition and defence of an ideal', in Gleeson, B. and Low, N. eds. *Governing for the Environment, Global Problems, Ethics and Democracy*, Basingstoke: Palgrave.

Thompson, J. (2009) *Intergenerational Justice: Rights and Responsibilities in an Intergenerational Polity*, London and New York: Routledge.

Tooze, A. (2021) *Shutdown, How Covid Shook the World's Economy*, Allen Lane.

Varma, S. (2012) 'Under Modi, tax burden has shifted from corporates to people', https://www.newsclick.in/under-modi-tax-burden-shifted-corporates-people, (accessed 25/01/2023).

VDem Gothenberg. (2022) *Democracy Report 2022: Autocracies Changing Nature*, Sweden: University of Gothenberg.

Waste, R.J. (1987) *Power and Pluralism in American Cities, Researching the Urban Laboratory*, New York: Greenwood Press.

Watson, G. (2023) 'Our government is being privatised by stealth', *The Age*, Melbourne 17/05/2023.

9
INEQUALITY AND POVERTY

Introduction

Thanks to the work of the World Inequality Lab hosted by the Paris School of Economics and the University of California, Berkeley, we now know a lot about the current state, history, and dynamics of inequality. The Lab, with some 40 researchers, works closely with the international World Inequality Database network, which brings together the work of over a hundred researchers on the five continents.

Thomas Piketty, who is a co-director of the Lab, has expressed something of the philosophy inspiring the research conducted by these organisations:

> Inequality is neither economic nor technological; it is ideological and political … In other words, the market and competition, profits and wages, capital and debt, skilled and unskilled workers, natives and aliens, tax havens and competitiveness – none of these things exist as such. All are social and historical constructs, which depend entirely on the legal, fiscal, educational, and political systems that people choose to adopt and the conceptual definitions they choose to work with.[1]

A crucial element of inequality is poverty. Poverty remains a blight on societies of both developed and developing economies. The 2023 Oxfam report on poverty links poverty with inequality, with economic institutions today providing soaring wealth to the few with mounting crises for the poorest people.[2]

Following the path of Karl Polanyi, we have tried in the foregoing chapters to plot the history of social and political transformations and the institutions that have resulted from them. In this chapter, we first consider inequality, then poverty,

DOI: 10.4324/9781003382133-11

and finally climate inequality. The question of how institutions and ideologies need to change is the subject of the following chapter.

Inequality since 1910

Let's look first at the historical data. Piketty's data and charts illustrate clearly what happened to the distribution of income in the 'Anglo-Saxon' countries (the USA, Canada, Britain, and Australia), compared with continental Europe and Japan (France, Sweden, Germany, and Japan). There is a significant difference between these two groups which stands out in the two charts in Figures 9.1 and 9.2.

At the start of the twentieth century, the top centile (top 1% income group of the population) received on average more than 20 per cent of total annual income. That distribution slid slowly after the First World War and fell sharply after the Second World War. Between about 1970 and 1980, the share of the top centile fell to below 5 per cent of the total in Australia and around 8 per cent even in the USA. Much the same occurred in continental Europe as in the Anglo-Saxon countries.

After 1980, however, there was a marked difference between these 'two worlds' (Anglo and European) with the share of the top centile rising to about 18 per cent of the total income in the USA, while in continental Europe the share remained below 12 per cent. Although, for obvious reasons, Piketty includes Australia in the

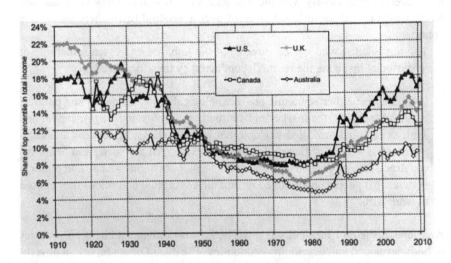

FIGURE 9.1   Income inequality in Anglo-Saxon countries, 1910–2010. Interpretation: The share of the top percentile in total income rose since the 1970s in all Anglo-Saxon countries, but with different magnitudes. Sources and series: see piketty.pse.ens.fr/capital21c.

(Source: http://piketty.pse.ens.fr/files/capital21c/en/pdf/F9.2.pdf).

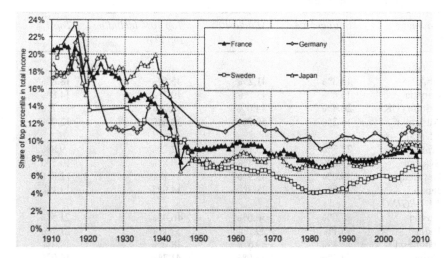

FIGURE 9.2   Income inequality: continental Europe and Japan, 1910–2010. Interpretation: As compared to Anglo-Saxon countries, the share of the top percentile barely increased since the 1970s.

(Source: http://piketty.pse.ens.fr/files/capital21c/en/pdf/F9.3.pdf).

Anglo Sphere, Australia's income inequality trajectory until 2008 was more like that of Europe, with the top centile share at just below 10 per cent in 2010. It has risen to 12.9 per cent since then (see Table 9.1).

When we look at the top thousandth in income receipt (the top 0.1%), the rise in the Anglo-Saxon countries is even more striking. From a level of between 2 per cent of total income in the 1970s, the US top thousandth rose to between 7 and 8 per cent in 2010. The British sector rose from under 2 per cent in the 1970s to between 5 and 6 per cent in 2010. Canada went from 2 per cent to 5 per cent, and Australia from 1 per cent to 3 per cent.[3]

Piketty charts income inequality in six 'emerging countries' (India, Indonesia, China, South Africa, Argentina, and Colombia) from 1910 to 2010. Though showing complex differences among the countries, especially between 1940 and 1950 (the period of war and decolonisation), the broad pattern of income inequality between 1980 and 2010 is similar to that found in America and Europe (Fig. 9.3).

Turning to the distribution of wealth or capital over a longer period, wealth inequality in Europe began from a much higher level than that of the USA and fell to a lower level than the USA after the two world wars. In Europe, the wealth of the top 10 per cent (decile) rose from 80 to 90 per cent of total wealth from 1810 to 1910, falling to 60 per cent by 1970. In the USA, the top decile owned just under 60 per cent of total capital in 1810, rising to about 80 per cent by 1910. After 1910, the share of the top decile fell to about 65 per cent by 1930, rising to 70 per cent by 2010.[4]

TABLE 9.1 Income inequality

Country	Bottom 50%	Middle 40%	Top 10% Including top 1%	Top 1%
ANGLO				
USA	13.3%	41.2%	45.5%	18.8%
Canada	15.6%	43.7%	40.7%	14.8%
Britain (UK)	20.4%	44.0%	35.7%	12.7%
Australia	16.2%	50.2%	33.6%	12.9%
EUR/JP				
France	22.7%	45.1%	32.2%	9.8%
Sweden	23.8%	45.4%	30.8%	10.5%
Germany	19.0%	43.9%	37.1%	12.8%
Japan	16.8%	38.3%	44.9%	13.1%
EMERGING				
Russia	17.0%	36.6%	46.4%	21.5%
India	13.1%	29.7%	57.1%	21.7%
Indonesia	12.4%	39.6%	48.0%	18.3%
China	14.4%	44.0%	41.7%	14.0%
South Africa	5.3%	28.2%	66.5%	21.9%
Argentina	16.2%	41%	42.8%	17.5%
Chile	10.2%	30.9%	58.9%	26.5%
Brazil	10.1%	31.4%	58.6%	26.6%

Source: The distribution of income in Anglo, European, and Emerging groups of nations has been tabulated from country figures in the *World Inequality Report 2022* pages 179–229. Country Sheets.

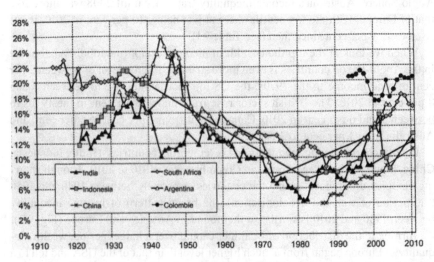

FIGURE 9.3 Income inequality in emerging countries, 1910–2010. Interpretation: Measured by the top percentile income share, income inequality rose in emerging countries since the 1980s, but ranks below US level in 2000 2010.

(Source: http://piketty.pse.ens.fr/files/capital21c/en/pdf/F9.9.pdf).

Inequality of capital ownership is historically always more marked and persistent than inequality of labour incomes. Piketty finds this regularity 'in all countries in all periods for which data are available, without exception, and the magnitude of the phenomenon is always quite striking'.[5]

The upper 10 per cent of those receiving income from labour in any country generally receive 20–35 per cent of total income from labour. The upper 10 per cent of those who receive income from capital always *own* more than 50 per cent of all capital – and in some societies at some periods up to 90 per cent. More striking is that 'the bottom 50 per cent of the wage distribution always receives a significant share of total labour income ... whereas the bottom 50 per cent of the wealth distribution owns nothing at all, or almost nothing'.[6]

Piketty has a vision of capitalism very similar to that of Karl Polanyi. Capitalism contains an antinomy – a law-like contradiction – that can only be resolved by embedding markets in social institutions designed to prevent capitalism from destroying itself and society.

Participation in the global economy, Piketty argues, is 'not a negative'.[7] The Asian countries in particular have benefited from openness to foreign influences, open markets for goods and services, and advantageous terms of trade. In this sense, globalisation has enabled the poor to catch up with the rich through the spread of technological knowledge, skill, and education both among and within nations. But this progress is conditional upon market economies embedded in stable, legitimate, and efficient governments guaranteeing high-quality education for all.

Yet capitalism also contains the potential for increased economic stratification and permanent class creation, in conditions when the growth rate of capital (growth of the stock market) is much higher than the growth rate of production (growth of the economy).

Classes of the Twentieth Century

In Europe, the two world wars and their public policy aftermath completely disrupted the patrimonial (inheritance-based) societies of the previous century in which the top decile of the wealthy owned most of what there was to own, and the least wealthy half of society owned virtually nothing.

Since around 1980, however, inequality has started growing worldwide, but most markedly in the USA and Britain. The classes Piketty discerns today are: (i) what remains of the 'hyperpatrimonial' class of pure (or almost pure) rentiers, (ii) the poor – typically a quarter of the population who own very little, if any wealth, and, between the two: (iii) the class of 'supermanagers', and (iv) a significant 'patrimonial middle class'.[8]

As noted, the 'class of rentiers', those for whom work for income is optional, suffered a sudden and catastrophic decline in the post-war world, but it still exists at the top of the wealth pyramid and has since recovered some of its ground. It is the least visible of classes, hiding its wealth and avoiding taxation in a wide range of tax havens. If it exerts political power, it is done for the most part discreetly. For the most part also, it hides conspicuous consumption, though there are exceptions.

A stratum that has emerged post-war, particularly in the USA, is the class of 'supermanagers' or, as its members like to think of themselves, a 'hyper-merito-cratic' class. The most senior executives who circulate amongst the boards and managements of the largest corporations can command 'skyrocketing pay pack-ages' 50 to 100 times the average wage.[9] As Piketty points out, there is very little evidence justifying such astronomical pay differentials on grounds of merit or pro-ductivity.[10] Rather, those at the top of the corporate scale, arriving there through luck, persistence, or inheritance, have acquired the power to determine their own remuneration with very little restraint and without end.

There is no clear demarcation between supermanagers and rentiers, the former transforming into the latter in so far as they do not consume all of their income.[11] Passing on their 'investments' to their progeny of course opens the door for their children to join the rentier class; a new kind of dynastic capitalism. Indeed, the *New York Times* reports that in the next ten years, 16 trillion US$ will be trans-ferred to the next generation. The largest share of that transfer will come from the (predominantly white) top 10 per cent .[12]

Finally, a significant 'patrimonial' middle class has emerged post-war, which Piketty argues 'was the principal structural transformation of the distribution of wealth in the developed countries in the twentieth century'.[13] Going back a cen-tury, the middle 40 per cent of the wealth distribution was almost as poor as the bottom 50 per cent. Today, Piketty observes, 'Tens of millions of individuals ... intermediate between rich and poor, own property worth hundreds of thousands of euros and collectively lay claim to one quarter to one third of national wealth'.[14]

World Inequality 2021

The World Inequality Lab now produces detailed and comprehensive reports on world inequality (2018, 2022). It is impossible here adequately to summarise these reports, nor to overstate their importance as the world's nations grapple with the future of their societies and democracies. They can be freely downloaded from https://wir2022.wid.world/. They are essential reading for social scientists, econo-mists, public officials, and political actors searching for a new social transforma-tion adequate to meet the challenge of climate change.

The *Foreword* to the report makes clear, if further clarification were needed, that the transformation to social democracy after World War II lasting from 1945 to 1980 was a period of shrinking inequality in many parts of the world, coupled with fast productivity growth and increasing prosperity, 'never matched since'. Thus, 'There is no prima facie evidence for the idea that fast growth demands or necessarily goes hand in hand with growing inequality'.[15]

Nevertheless, the neoliberal regression from the 1980s debated and planned by the Mont Pelèrin Society and promulgated over 40 years by the many, well-funded 'second hand dealers in ideas' (to cite Hayek) was triggered by the 'panicked reaction to the slowdown of growth in US and UK in the 1970s'. The problem of 'slowdown' was attributed to the institutions that kept inequality low: minimum wage, union militancy, progressive taxes, regulation, and planning. What was needed was an entrepreneurial culture that celebrated the unabashed accumulation of private wealth.

The Reagan–Thatcher transformation, the authors of the Foreword insist, was:

> The starting point of a dizzying rise in inequality within countries that continues to this day. When state control was (successfully) loosened in countries like China and India to allow private sector-led growth, the same ideology got trotted out to justify not worrying about inequality, with the consequence that India is now among the most unequal countries in the world (based on this report) and China risks getting there soon.[16]

There is also a paucity of official government data. Economic growth figures are collected and published each year. Yet these figures tell us nothing about how growth of income and wealth is distributed across the population. Nor do official figures record race and gender inequalities, or the distribution of climate impacts and costs. Knowing these figures is critically important for democracy. Nevertheless, the figures from the World Inequality Lab provide a reasonable indicator.[17]

Globally in 2021, the average adult earned US$23,380 per year and owned US$102,600. But there were huge disparities between and within countries. The richest 10 per cent of the global population take home 52 per cent of world income compared with the poorest 50 per cent who earn 8.5 per cent. Wealth inequality is even more pronounced. The richest 10 per cent own 76 per cent of all wealth. The poorest half of the world's people own just 2 per cent of total wealth. Regionally, the least income inequality is found in Europe, and the greatest in the Middle East and North Africa:

> Inequality varies significantly between the most equal region (Europe) and the most unequal (Middle East and North Africa: MENA). In Europe, the top 10% income share is around 36%, whereas in MENA it reaches 58%. In between these two levels, we see a diversity of patterns. In East Asia, the top 10% makes 43% of total income and in Latin America, 55%.[18]

Figures for the distribution of income and wealth in Anglo, European, and Emerging groups of nations are tabulated from country figures in the World Inequality Report 2022 pages 179–229. These groupings (in Tables 9.1 and 9.2) correspond with the historical figures for income inequality charted by Piketty shown above (in Figures 9.1, 9.2, and 9.3) with the addition of Russia, and in place of Colombia, Chile, and Brazil.

Among the emerging nations group, there is a distinct subgroup of the very poor in wealth. In these nations – South Africa, Chile, Brazil, Poland, and Mexico

TABLE 9.2 Wealth inequality

Country	Bottom 50%	Middle 40%	Top 10% Including top 1%	Top 1%
ANGLO				
USA	1.5%	27.8%	70.7%	34.9%
Canada	5.8%	36.5%	57.7%	25.0%
Britain (UK)	4.6%	38.2%	57.1%	21.3%
Australia	6.1%	37.7%	56.2%	23.7%
EUR/JP				
France	4.9%	35.6%	59.5%	27.0%
Sweden	5.8%	36.2%	58.0%	27.3%
Germany	3.4%	37.1%	59.6%	29.7%
Japan	5.8%	36.5%	57.8%	24.5%
EMERGING				
Russia	3.1%	22.8%	74.1%	47.7%
India	5.9%	29.5%	64.6%	33.0%
Indonesia	5.5%	34.3%	60.2%	29.4%
China	6.4%	25.8%	67.8%	30.5%
South Africa	-2.4%	16.8%	85.7%	55.0%
Argentina	5.7%	36.1%	58.2%	25.7%
Chile	-0.6%	20.1%	80.4%	49.6%
Brazil	-0.4%	20.6%	79.8%	48.9%

Source: The distribution of wealth in Anglo, European, and Emerging groups of nations has been tabulated from country figures in the *World Inequality Report 2022*, pp. 179–229.

– the bottom 50 per cent own a minus quantity, that is, they *owe* more than they *own* in wealth. There are of course many extremely poor nations not listed in the World Inequality Report. The USA figure of (plus) 1.5 per cent barely scrapes into positive territory and far below that of other 'Anglo' countries.

The Report states that global inequalities are close to 'early twentieth century levels, at the peak of Western imperialism'. While global inequalities between countries have declined over the past two decades, inequality *within* most countries has increased. 'The gap between the average incomes of the top 10% and the bottom 50% of individuals within countries has almost doubled'.[19]

Wealth inequalities have grown for the top 1 per cent of the distribution. This top tier took 38 per cent of all additional wealth generated since the mid-1990s, while the bottom 50 per cent took just 2 per cent. Billionaires in 2021 held 3.5 per cent of all household wealth.[20] This gross inequality stems from 'inequality in growth rates between the top and the bottom segments of the wealth distribution. The wealth of the richest individuals on earth has grown at 6% to 9% per year since 1995, whereas average wealth has grown at 3.2% per year'.[21]

As argued in Chapter 8, the success of the social purpose of democracy depends on the capacity of the state to implement the 'march to equality': to enact policies for the majority, for social justice, better working conditions, and the defeat of poverty, squalor, and ignorance. In this respect, the Report finds that

> over the past 40 years, countries have become significantly richer, but their governments have become significantly poorer. The share of wealth held by public actors is close to zero or negative in rich countries, meaning that the totality of wealth is in private hands.[22]

That trend was magnified by the Covid crisis during which governments world-wide borrowed the equivalent of 10–20 per cent of the Gross Domestic Product – essentially from the private sector. Thus, 'the low wealth of governments has important implications for state capacities to tackle inequality in the future, as well as the key challenges of the 21st century such as climate change'.[23]

The Report points to entrenched gender inequality. Globally, women's share of total income from work (labour income) stands at less than 35 per cent of total labour income – up from 30 per cent in 1990. But in a gender-equal world, women, half the world's population, would earn close to half of total income. Progress has been slow globally, and, 'dynamics have been different across countries, with some recording progress but others seeing reductions in women's share of earnings'.[24]

Poverty

Dividing populations in terms of income and wealth implies, indirectly, a measure of poverty. Yet poverty means more than inequality. Poverty concerns the life experience of those at the bottom levels of income and wealth. For instance, the Joseph Rowntree Foundation UK Report for 2023 stated, 'This winter, 7.2 million low-income households are going without essentials – hungry, cold, without basics like showers, toiletries or adequate clothing – and 4.7 million are behind on their bills'.[25]

Poverty is normally measured as 'relative poverty': the percentage of the population falling below one half of a particular country's median household income. But this method masks the very different life and death experiences of the poor among different nations. There is no consistent, annually reported, comparative measure of these experiences among nations (absolute poverty).

The UN Sustainable Development Goal Number One is 'End poverty in all its forms'. The primary focus of UN investigations is 'extreme poverty'. Thus,

> Between 2015 and 2018, global poverty continued its historical decline, with the extreme poverty rate falling from 10.1 per cent to 8.6 per cent. This means that the number of people living on less than US\$1.90 a day dropped from 740 million to 656 million over this period.[26]

Nevertheless, the report continues, Covid-19 has made a severe dent in that progress. Poverty increased sharply from 2019 to 2020, the first rise since 1998 and the largest since 1990.

The Oxfam Report (2023) on the international situation states,

> We are living through an unprecedented moment of multiple crises. Tens of millions more people are facing hunger. Hundreds of millions more face impossible rises in the cost of basic goods or heating their homes. Climate breakdown is crippling economies and seeing droughts, cyclones and floods force people to flee their homes.[27]

In the USA, the rate of poverty is the most extreme in the developed world. A report published in 2004 argued that the USA is exceptional for the high degree of poverty in that country.[28] The report examines in detail the reasons for the exceptional situation. Another study published in 2003 showed that 'cumulative poverty over many years is the fourth leading cause of death in the United States'.[29] Table 9.3 derived from OECD figures and updated to 2020 shows the

TABLE 9.3 Poverty in OECD countries 2019 and 2020–2021

Country	Overall 2019	Overall 2020/21	Children 2019	Poverty gap 2019
ANGLO				
USA	17.8%	15.1%	20.9%	39.8%
Canada	12.4%	8.6%	14.2%	30.4%
Australia	12.1%	12.6%	12.5%	28.7%
Britain (UK)	11.1%	11.2%	11.8%	35.5%
New Zealand	10.9%	12.4%	14.1	26.2%
EUROPE				
Germany	10.4%	10.9%	12.3%	26.5%
Poland	10.3%	9.8%	9.3%	28.4%
Hungary	10.1%	9.2%	11.8%	29.2%
Austria	9.8%	10%	11.5%	35.4%
Sweden	9.3%	8.8%	9.3%	22.5%
France	8.3%	8.4%	11.5%	23.9%
Finland	6.3%	5.7%	3.6%	21.0%
Denmark	5.5%	6.5%	2.9%	31%
ASIA				
South Korea	17.4%	15.3%	14.5%	35.5%
Japan	15.7%	15.7%	13.9%	33.7%
Average of 25 OECD countries Not including USA	10.7%	11.7%	20.9%	39.8%

Sources: OECD *Confronting Poverty* 'Fact 4' https://confrontingpoverty.org/poverty-facts-and -myths/americas-poor-are-worse-off-than-elsewhere/ 2020–2021 rates: https://povertyandinequality .acoss.org.au/poverty/poverty-rates-in-oecd-countries/ accessed 29/03/2023.

percentage of people falling below 50 per cent of the country's median household income (relative poverty).[30]

From the figures, it is clear that relative poverty remains persistent. However, rates vary considerably among OECD nations, with the Anglo countries registering more poverty than those of continental Europe. Scandinavian countries have the lowest poverty rates, though Sweden is somewhat higher than others (possibly because of its higher migration intake). Formerly eastern bloc countries, including Germany (only fully reunified since 1991), have higher relative poverty rates. Japan and Korea have poverty rates comparable to those of the USA.

The World Inequality Report also addresses 'carbon inequality'. Global income and wealth inequalities, the Report argues, are tightly connected to ecological inequalities and to inequalities in contributions to climate change. The World Inequality Lab has issued a separate report focused on the inequalities of climate change.[31]

Climate Inequality

On 19 March 2023, the Intergovernmental Panel on Climate Change (IPCC) published its sixth assessment report. The report asserts with impeccable evidence and 'high confidence' that 'Human activities, principally through emissions of greenhouse gases, have unequivocally caused global warming, with global surface temperature reaching 1.1°C above 1850–1900 [levels] in 2011–2020'.[32]

The IPCC AR6 Synthesis Headline Statements explicitly link climate change with inequality. 'Global greenhouse gas emissions have continued to increase, with unequal historical and ongoing contributions arising from unsustainable

TABLE 9.4 Global carbon emissions 1850 to 2019

Date	Global emissions (billion tonnes)	Emissions per capita (tonnes per person)
1850	1.0	0.8
1880	2.5	1.8
1900	4.2	2.7
1920	6.6	3.5
1950	10.9	4.3
1980	30.2	6.8
2000	35.3	5.8
2019	50.1	6.6

Source: *World Inequality Report* Table 6.1 Global carbon emissions, 1850–2019, p. 118.

Interpretation: Emissions of carbon dioxide equivalent (including all gases) from human activities (including deforestation and land-use change).

energy use, land use and land-use change, lifestyles and patterns of consumption and production across regions, between and within countries, and among individuals'.[33]

Action on climate change requires social progress 'for all'. The IPCC Report continues: 'Climate resilient development integrates adaptation and mitigation to advance sustainable development for all and is enabled by increased international cooperation including improved access to adequate financial resources, particularly for vulnerable regions, sectors and groups, and inclusive governance and coordinated policies'.[34] The report soberly warns that what the world's nations and international regimes decide to do in the next ten years will have impacts now and for thousands of years into the future.

In recognition of the immense importance of the threat of global heating, the World Inequality Lab published a separate report in 2023: *Climate Inequality Report, Fair Taxes for a Sustainable Future in the Global South*. The question of policy – 'fair taxes' and institutional change – will be discussed in the next chapter. Here we look at the key issue which is truly the (invisible to policy makers) elephant in the room.

In summary[35]:

- The climate crisis is fuelled by the polluting activities of a small fraction of the world's population. The global top 10 per cent are responsible for almost half of the total carbon emissions. The top 1 per cent are responsible for more emissions than the whole of the bottom 50 per cent.
- Inequality of emissions within countries now accounts for about one-third of the total (an almost complete reversal since 1990).
- Reducing individual carbon budgets at the top can free up the required budget to lift people out of poverty.
- Poverty and vulnerability to climate hazards (weather related) mutually reinforce one another. Low-income regions face agricultural losses of 30% and more. 780 million people across the world face the combined risk of poverty and serious flooding.

Water scarcity is also a major threat. A UN report records that 2.3 billion people live in water-stressed countries, of which 733 million are critically water stressed. 'Water can be scarce for many reasons: demand for water may be exceeding supply, water infrastructure may be inadequate, or institutions may be failing to balance everyone's needs'.[36]

Global *ecological* inequality 'takes many forms, including inequality of access to natural resources, inequality of exposure to pollution and to the catastrophes induced by unsustainable use of these resources, and inequality of contribution to environmental degradation'.[37]

In what follows, there is space only for the bare bones of the full and complex treatment of carbon inequality in the two *World Inequality* reports. The figures themselves tell the story of the growth of carbon pollution in the industrial age, the unequal contribution to carbon pollution and global heating among socio-economic divisions, inequality among world regions, and inequality among nations and among the socio-economic divisions within each. Finally, we consider the carbon budget required to meet containment of global heating to an increase from pre-industrial times of 1.5°C and 2.0°C.

Carbon Inequality among Socio-economic Divisions

Responsibility for carbon emissions is extremely unequally shared among the socio-economic divisions of the global population (Table 9.5). The bottom 50 per cent of the population accounted for a mere 1.6 per cent of total emissions in 2019, compared with 24 per cent for the top 10 per cent and 1.7 per cent of total emissions for the top 1 per cent – more than for the entire bottom 50 per cent.

It is very clear that in all countries the contribution to climate change from the top 1 per cent of the population is many times greater than that of the bottom 50 per cent. Growth of emissions between 1990 and 2019 has been overwhelmingly skewed to the top of the socio-economic ladder, with the greatest growth (168 per cent between 1990 and 2019) in the top 0.001 per cent (or top ten thousandth of the population).

Carbon Inequality among Regions

There is a difference between the emissions per person generated solely within a region's territory and the total per-person emissions including those associated with the importation of goods and services from the rest of the world. In Table 9.6, the first are termed 'territorial emissions', and the latter 'footprint emissions'.

TABLE 9.5 Inequality among socio-economic divisions 1990–2019

Socio-economic divisions	Per capita emissions (Tonnes CO_2e per person)		Total emissions (billion tonnes CO_2e)		Growth per capita emissions (1990–2019)	Growth in total emissions (1990–2019)	Share in emissions growth (1990–2019)
	1990	2019	1990	2019			
World population	6.2	6.6	32	50.5	7%	58%	100%
Bottom 50%	1.2	1.6	3.1	6.1	32%	96%	16%
Middle 40%	6.0	6.6	13.3	20.4	4%	54%	39%
Top 10%	30	31	15.7	24	4%	54%	45%
Top 1.0%	87	110	4.5	8.5	26%	87%	21%
Top 0.01%	323	467	1.7	3.6	45%	114%	10%
Top 0.001%	1,397.0	2,531	0.7	2	81%	168%	7%

Source: *World Inequality Report,* p. 124, Table 6.6 Emissions growth and inequality, 1990–2019.

TABLE 9.6 Average footprint and territorial per capita emissions by world regions

Region	Footprint: tonnes of CO_2e per person per year	Territorial: emissions of tonnes of CO_2e per person per year
World	6.6	6.6
Sustainable Level <1.5 degrees	1.1	
Sub-Saharan Africa	1.6	2.1
South and South-East Asia	2.6	2.7
Sustainable Level <2.0 degrees	3.4	
Latin America	4.8	4.9
MENA	7.4	8.0
East Asia	8.6	9.4
Europe	9.7	7.9
Russia and Central Asia	9.9	11.9
North America	20.8	19.8

Source: Figures 6.3a and 6.4, *World Inequality Report*, p. 119.

Interpretation: Values include emissions from domestic consumption, public and private investments, as well as imports and exports of carbon embedded in goods and services traded with the rest of the world. The sustainable level corresponds to an egalitarian distribution of the remaining carbon budget until 2050.

For example, when North Americans import smartphones from East Asia, carbon emissions generated in the production, transport, and sale of these phones are attributed to the *footprint* of North Americans and not to East Asians. The 'footprint' is the most informative way of measuring emissions associated with the standard of living of individuals across the world.

Territorial emissions are routinely used by national agencies to report progress on emissions reduction for the purpose of international climate agreements. But high-income countries can reduce their reported territorial emissions while externalising to the rest of the world the carbon-intensive industries producing goods for home consumption (which they then import). Factoring in the carbon embedded in imported goods and services adds to the climate change mitigation efforts required of high-income countries, especially in Europe where imports represent a notable share of per capita emissions.

Carbon Inequality within and among Nations

The World Inequality Report includes individual reports on 26 individual nations in 'Country Sheets' (pp. 178–229). These are only a small sample of the world's nations, but they include all the nations with the highest rate of carbon emissions. Here the distribution of emissions from different socio-economic groups can be compared with those countries' historical data for inequality of income and wealth already shown in Tables 9.1 and 9.2.

TABLE 9.7 Carbon inequality within and among nations

Country	Bottom 50%	Middle 40%	Top 10% Including top 1%	Top 1%	Total per capita
ANGLO					
USA	9.7	22.0	74.7	269.3	21.1
Canada	10.0	20.9	60.3	190.2	19.4
Britain (UK)	5.6	10.9	27.7	76.6	9.9
Australia	9.7	21.8	60.2	196.0	19.6
EUR/JP					
France	5.0	9.3	24.7	77.5	8.7
Sweden	5.4	10.1	27.9	97.3	9.5
Germany	5.9	12.2	34.1	117.8	11.3
Japan	6.3	12.4	37.9	109.2	11.9
EMERGING					
Russia	6.8	11.7	41.7	186.1	12.3
India	1.0	2.0	8.8	32.4	2.2
Indonesia	1.4	3.5	11.8	42.2	3.3
China	3.0	7.2	36.4	138.9	8.0
South Africa	3.0	6.5	31.3	116.4	7.2
Argentina	3.5	7.0	19.0	58.0	6.5
Chile	2.7	5.8	26.1	108.2	6.3
Brazil	2.2	4.3	17.7	72.1	4.6

Source: 'Carbon Tables' in 'Country Sheets' pp. 178-229, *World Inequality Report*.
Note: Average greenhouse gas footprint: tonnes CO_2e per capita 2019 for socio-economic groups in income and wealth.

Carbon Budgets

There is a limited amount of greenhouse gases the world can afford to release into the atmosphere before the average global temperature rises to dangerous levels. These levels indicate tipping points beyond which it becomes extremely expensive to stop further global heating (at 1.5 degrees above pre-industrial levels). Beyond this level, global heating feeds on itself, laying waste to natural ecosystems and generating even more heating (at 2.0 degrees above pre-industrial).

These limits give us 'carbon budgets' for future expenditure of our use of the atmosphere as a carbon sink. Between 1850 and 2020, 2,450 billion tonnes of greenhouse gas were sent into the atmosphere (using it as a carbon sink). According to the 2022 IPCC report, to stay below 1.5 degrees of heating, we can afford to send into the atmosphere another 300 billion tonnes. If we set the limit at 2 degrees, we can afford to emit another 900 billion tonnes. But 2 degrees of heating puts us in extremely dangerous territory.[38]

This is why the Secretary General of the United Nations is calling for urgent and immediate action. This must mean stopping any expansion of existing oil and gas reserves, shifting subsidies from fossil fuels to a just energy transition, and establishing a global phase-down of existing oil and gas production compatible with the 2050 global net-zero target[39].

At current global emissions rates, the 1.5°C budget will be depleted in six years and the 2°C budget in 18 years. We must now make the carbon budget personal and individual.

> Sharing the remaining carbon budget to have an 83 per cent chance to stay below 1.5°C global temperature increase implies an annual per capita emission level of 1.1 tonnes per person per year between 2021 and 2050 (and zero thereafter). Sharing this same budget between 2021 and 2100 implies per capita annual emissions of 0.4 tonne.[40]

This is about six times less than the current global average.[41]

Not all nations have historically generated global heating through carbon emissions. The old industrial world contributed much more than the emerging industrial nations even while global heating was being explored scientifically. If we take historical responsibilities into account, high-income nations have no carbon budget left at all.[42]

Inequality is not inevitable. It is a political choice.[43] Poverty in rich countries is a political choice. The post-war transformation to social democracy discussed in Chapter 3 (above) demonstrates how greater equality was achieved in Britain and under what circumstances. In the post-war years, the evil of poverty was to be fought and eliminated.

The political choice, however, is not a simple one. Ideology and its concomitant institutions are bound up with political choices, always against the background of what we might call the political ethos of a nation's population. By 'political ethos' I want to imply something ontological more than sociological: what it is felt to *be* a citizen of a nation, who and how some occupants are excluded from that way of being, and how that sense of being is embedded in the nation's history. To illustrate the point, we turn to a country with which I, the author, am most familiar: Australia.

Ideology, Poverty, and Political Ethos: Australia

Australia provides an instance of the way ideology, institutions, and government failure exacerbate inequality and poverty against the background of the political ethos of a rich nation.

Australia's wealth is based both on the ingenuity of its people and the immense agricultural and mineral resources within its bounds: 7.688 million sq. km. The world's fifth largest national area after Russia, Canada, the USA, and Brazil.[44] Its population density is tiny: 3.3 people per sq. km., compared with Russia (9), Canada (4), USA (34), and Brazil (25). Typically, international investment in its resources is huge, but investment in Australian innovative technology has been weak and sometimes non-existent. The service sector occupies 80 per cent of its economy,[45] yet within that sector, the 'care economy' – (social care with 1.8 million employees, education and training with over a million, and the health sector with over 600,000[46]) is seriously underfunded.

Not only does Australia have one of the world's largest per capita carbon emission footprints, it also exports prodigious quantities of fossil fuels (coal and gas). For ten years between 2013 and 2022, the reality of climate change was denied or obfuscated. Attempts to impose emission controls were politically vilified during the so-called 'carbon policy wars'.

The Labor government elected in 2022 has taken cautious steps to bring the country's carbon footprint down. Importantly, the Greens Party have pitched themselves as a party of government rather than opposition. There have been constructive negotiations between the Greens and Labor on matters of substance. The key question of Australia's export of fossil fuels, however, remains unanswered.

It might be expected that a country with such resources would not allow poverty among its people. But, since most of these mineral resources (including even water rights) are leased under contract to international investors, Australians derive much less benefit from the sale of these resources than do the international investors.

Australia has one of the world's most advanced democratic constitutions. In a bicameral federal system, voting (at both federal and State levels) is compulsory with preferential voting in the lower house and proportional representation in the upper house.[47] But the capacity of the public service has been greatly weakened since the neoliberal regression in the 1980s by privatisation of key public utilities at State level and outsourcing of many public advice functions to consultants.

Successive governments of both main parties have, since the 1980s, downsized and outsourced the public service. Many tenured senior positions have been replaced by consultants with a vested interest in serving the priorities of politicians. The concept of frank and fearless advice by the public service to the political level has been abandoned.

The gradual deterioration of the public service has empowered politicians at the expense of independent professional advice and erased the long-term memory of the results and conditions of public policy governance to the extent of 'political amnesia' (in the words of the former editor of the Australian Financial Review, and now chief political correspondent of the Australian Broadcasting Commission, Laura Tingle[48]).

The failures of inequality follow from the reaction to the Covid-19 pandemic, the subsequent inflation, and the egregious treatment of the unemployed. The ideology is that of the neoliberal regression. The institutions are the independence of the central bank and the way government funding is provided to support workers in periods of unemployment. The failure is the intentional maintenance of poverty within a significant section of the population at the bottom end of income and wealth.

Poverty in Australia

The Australian Council of Social Service (ACOSS) with the University of New South Wales carries out regular surveys of poverty and inequality. In this context, poverty does not mean starvation as it can do in poor countries. ACOSS defines poverty as a relative concept used to describe the people in a society that cannot participate in the activities that most people take for granted. That is to say: 'enough money or resources for the basic needs of life – enough food for oneself and for one's family; a roof over one's head; and resources to cover clothing, education and health expenses'.[49]

The surveys use two international poverty lines to measure poverty in Australia: 50 per cent and 60 per cent of median income. People living below these lines are regarded as living in poverty. Income for most assessments excludes housing costs. Households with lower housing costs are able to afford a higher standard of living than those on the same income with higher housing costs.

The 2022 ACOSS survey found that there were '3.3 million people (13.4%) living below the poverty line of 50% of median income, including 761,000 children (16.6%). In dollar figures, the poverty line works out to AU$489 a week for a single adult and AU$1,027 a week for a couple with 2 children'.[50] Many of those have an income as much as AU$304 per week *below* the poverty line, described as deep poverty.

First Nations people (Aboriginal and Torres Strait Islander people) are among the poorest Australians. 31.4 per cent were living below the (50%) poverty line in 2016, declining slightly from 33.9 per cent in 2006. The poverty rate was much higher in very remote areas – rising to above 50 per cent in 2016 – than in cities and regional towns.[51]

Most people living in poverty are dependent on social security (government income support payments). The majority are renting (56%). Sole parent households have the highest poverty rate. Children in sole parent families are more than three times more likely to live in poverty (44%) than children in couple families (13).

Economic conditions and fluctuations in government payments have been the main cause of changes in overall poverty. Among the richer OECD countries (according to a 2019 OECD report), in 2016 Australia (with 12.1%) was part of an English-speaking group, which had poverty rates above the OECD average of 11.8 per cent. However, the USA, with 17.8 per cent, had the second highest rate overall.[52]

The Australian Political Ethos

Ideology, institutions, and governance failure are interwoven in Australia. But they have to be understood against the backdrop of the national political ethos.

The political ethos of Australia is that of a country formed by immigration, following frontier wars to defeat and subjugate the indigenous population. Australians are a generous and compassionate people who don't hesitate to help those they recognise as their fellow citizens. Many instances of such unwavering voluntary mutual help occurred during the fires of 2019 and the floods of 2022. Australia has a strong record of volunteering: in its rural fire service, in its beach life-saving service, and in the health and aged care sectors.

But the Australian ethos is also shaped by stereotypes of those who fall out of the category of 'citizens like us'. Those unrecognised include three groups. The most impoverished and alienated among indigenous people, poor recent migrants, whose only avenue for escape from intolerable conditions in their country of origin is to turn to people smugglers offering rides in overcrowded and unseaworthy boats, and the unemployed.

It should be noted, however, that once people, whatever their background, are seen as actual citizens with a valued place in the community as opposed to generalised abstractions, attitudes change dramatically. Those people become valued and supported. A striking example is the way in which migrants from Sri Lanka, Africa, and the Middle East have been absorbed into rural communities who come to embrace them as their own. The unemployed and otherwise impoverished are widely supported by volunteer agencies supplying basic needs, particularly food and some basic shelter.

In talking about the political ethos of a country, there is no implication of a homogeneous opinion. There are many shades of opinion in Australia about all three of the above groups. The political ethos is about the spectrum of attitudes that shape political decisions in a democracy. Australians will at least tolerate harsh treatment of boat people and the unemployed and paternalistic treatment of the indigenous population. Political leaders feel they have to be aware of the spectrum of attitudes and not step too far outside its parameters.

The Australian Left has from the late nineteenth century emphasised protecting the wages and conditions of the working class. In the post-war period, the Left approach to social welfare can be summarised as 'the workers' welfare state. From the 1980s, the rhetoric changed to prioritise Australian 'working families' and even 'aspirational' workers – those entrepreneurs and small business owners who aimed to build wealth. Exclusionary political rhetoric demonising economic or racial minorities can find acceptance amongst the working and middle classes.

The political ethos evolving since the 1980s is very different from that prevailing during and immediately following World War II, when a widespread feeling of compassion prevailed, as in Britain, for those who had given their lives to defend their country and its democracy, for the suffering during the depression, and for the families and children of those who had suffered. Since the 1980s, the ethos has become one of comfortable individualism with awareness of suffering driven to the margins.

The Central Bank, Inflation, and Unemployment

The Australian Parliament in 1959 gave the Commonwealth Bank, a government-owned entity, the function of a central or 'reserve' bank. The Reserve Bank of Australia was separated from the commercial functions of the Commonwealth Bank, which became the Commonwealth Banking Corporation.[53] The bank was fully privatised in 1996.

The Reserve Bank has three stated functions: maintaining the stability of the Australian currency, maintaining full employment, and providing for the economic prosperity and welfare of the people of Australia.[54] There was a major ideological change in the balance of these functions following Australia's embrace of the neoliberal regression in 1983. Following a gradual shift from government to market control of banks, 'the Australian financial landscape was transformed to a virtually fully deregulated system'.[55] An important ideological change occurred, which affected the definition of full employment. As we saw in Chapter 4, Margaret Thatcher discovered that the most effective way to prevent the British trades unions from demanding and obtaining higher wages was through the threat of unemployment.

Initially, the Australian Labor Government from 1983 sought to contain wage growth (and hence the threat of a wage-price spiral) through an 'Accord' with the unions. The unions agreed to restrict wage demands and the government pledged to minimise inflation and increase the 'social wage'. The latter included a National Occupational Health and Safety Commission, tax cuts for low and middle-income workers, increased pension and unemployment benefits, family income supplements for low-income families, and mandated 3 per cent (of wages) superannuation.[56] However, the Accord also included a reduction of the top personal income tax rate from 60 to 47 per cent.

Subsequently, 'full employment' was redefined by the Reserve Bank. The first step was to argue, reasonably, that there would always be some 'natural' unemployment as people left jobs and sought new ones. As the Reserve Bank came to define its primary role as containing inflation and compensating for recession by adjusting interest rates, a new concept of 'natural' unemployment was devised.

This concept was called, impenetrably, the 'Non-Accelerating Inflation Rate of Unemployment' (NAIRU). What that means is the inflation rate beyond which interest rates have to be raised in order to generate higher unemployment. Thus, it is a disciplinary measure to hold down wage demands and stop an inflationary wage-price spiral. The actual level of NAIRU is not precisely defined, but is believed to lie between 3 and 6 per cent of the adult workforce.

While business lobbies were arguing furiously against an increase in the basic wage of 5.75 per cent for the year 2023–2024, a report for the Governance Institute of Australia found that in the financial year 2022–2023 CEOs of ASX (Australian stock exchange) listed companies received a 'base' pay rise of 15 per cent. The heads of the top 200 listed companies received an average fixed pay increase of 19 per cent. CEOs of non-listed companies and not-for-profit companies received an average of between 8 and 9 per cent.[57] While the members of this class of 'super-managers' is small in number, the top 1 per cent of income earners took 12.9 per cent of national income, plainly adding to inflation.

An OECD publication released on 7 June 2023 reports that the inflation in Australia that accelerated in 2022 was caused mainly by a surge in company profits.[58] The president of the European Central Bank, Christine Lagarde, 'conceded to a parliamentary committee that there appeared to be cases of increased corporate profits making the European Union's inflation worse'.[59] On the same day, the governor of Australia's Reserve Bank offered advice to people struggling with increasing interest rates: 'If people can cut back spending, or in some cases find additional hours of work, that would put them back into a positive cash flow position'.[60] Many would like to work more, but the Bank's interest rate increases are explicitly intended to increase unemployment.

The position of the Reserve Bank is not neutral but ideologically biased. As of 2023, the actions of the Reserve Bank to reduce inflation are both creating unemployment and contributing to inflation. The top 1 per cent are exempted from consideration. Increased interest rates add to the cost to landlords of the borrowed money invested in rental housing, which is then passed on to renters. A former governor of the Reserve Bank, Bernie Fraser, argues that 'neoliberal policies have undermined the bank's attempts to improve the welfare of all Australians'. He says that those who come out on top are 'the well-heeled and the people who are able to take advantage of market power, and so on, and it's the people at the other end, the workers for the most part, who are left to pick up the crumbs'.[61]

Robodebt

For many years since the 1980s, the RBA strategy was, in its own terms, a huge success. Wage inflation was stamped out, and eventually incomes stagnated after the Great Recession of 2008. But this success came with rising costs to the government budget in unemployment claims. Governments of both main parties since the 1990s have refused to raise unemployment benefits to keep pace with inflation.

In 2016, the (Liberal–National) Coalition government introduced a new method of calculating 'overpayments' made to the unemployed. Instead of manually calculating overpayments on the basis of each separate period of unemployment, the new method used a computer algorithm to compare payments made with averaged income data obtained from the Australian Tax Office. This attempt to save money for the budget became known as 'Robodebt'.

Many of those who had suffered periods of unemployment had their suffering compounded as debt notices, sometimes for thousands of dollars, were automatically sent to them in some cases years after they had found employment. The scheme presumed guilt and demanded that the accused prove their innocence. At least three suicides have been linked to the scheme which was eventually found to be illegal from the start. In 2019, the government settled a claim for AU$1.8 billion made on behalf of the victims of the scheme.[62]

The Labor Government elected in May 2022 set up a Royal Commission to hear from the victims of Robodebt and investigate who was responsible for the scheme. A major concern of the Commission has quite properly been to establish who knew what about the legality or otherwise of Robodebt. But what has become clear is that the question of strict legality is overshadowed by the glaringly obvious injustice and impropriety of the scheme. Anyone who had even passing knowledge of the scheme could see that averaging annual income data could not validly be used to assess overpayments made on the basis of need during specific time periods.

The Commission reported in July 2023.[63] Chaired by Catherine Holmes SC, formerly chief justice of Queensland's supreme court, the Commission found that four senior government ministers, including the former Prime Minister (Scott Morrison), and senior public servants failed to govern within the law. Morrison, who initiated the scheme while Minister for Social Services, the Commission found, failed to meet his responsibility to ensure that the scheme was lawful. Public servants failed to serve the public and chose instead to serve the political executive.

The Report contains a sealed section (not available to the public, to ensure that subsequent prosecutions are not compromised) with specific charges against individuals to be referred for further action by authorities including the federal police.

Gareth Hutchens, business reporter for the ABC, traces the origin of Robodebt to the 'dole bludger' narrative promoted in the 1970s:

> To sell newspapers or to gain political advantage people in a position to influence members of the public have played on their fears that they, as hard workers and substantial taxpayers, were being sponged on by lazy ne'er-do-wells or scheming cheats.[64]

There is no evidence that this stereotype has any basis in truth. It is manufactured to support a punitive government approach to employment policy.

Political journalist, Waleed Aly, captures the underlying issue. He asks: 'Why is this not part of a bigger scandal?' He continues: 'If we're honest, the unemployed aren't treated as part of our political community'.[65] The vilification of the unemployed, culminating in Robodebt, is a demonstration of Australia's political ethos.

Australia's empathy for those unrecognised as 'like us' was tested again late in 2023, when the Labor government puts a referendum to the people, amending the constitution to recognise Australia's indigenous people and give them a permanent and representative 'Voice' to parliament. The referendum was defeated with 60% of Australians voting 'No'.

Covid-19

The Australian Government's response to the Covid-19 pandemic from 2019 to 2022 was rapid, dramatic, inconsistent, and fundamentally flawed. Like many governments around the world, the initial reaction was to stop people from infecting one another by moving around and gathering in close proximity. This approach was highly effective, but it was a stopgap measure while vaccines were developed and distributed. The anticipated effect was severe and immediate recession as businesses closed and people stopped buying.

The progress and impact of the 'Coronacession' was chronicled in detail in 2021 by historian Adam Tooze. He rightly observes that 'What is at stake in the response to pandemic threats is not just a vast amount of economic value. What is at stake are basic questions of social order and political legitimacy'. He argues that political systems around the world were – and still are – ill-prepared for pandemic threats.[66]

The Reserve Bank used its singular power to stimulate the economy by reducing interest rates to near zero. The Australian Government also employed unheard of fiscal measures to manage the recession. The Government committed AU$291 billion in direct economic support up to May 2021. A large proportion of this support went to businesses and not-for-profit organisations to help them keep staff on the payroll and paid, even while they were not working. These businesses were able to receive AU$1,500 (before tax) per fortnight per employee to cover the cost of wages.[67] Unemployment benefits were also doubled.

15,672 people had died with or from Covid-19 up to 31 January 2023.[68] The impact of Covid exacerbated inequality. The number of people who died of the virus is 'three times higher among the "least advantaged" members of society than the "least disadvantaged" according to the latest figures from the Australian Bureau of Statistics'.[69] The mortality rate for First Nations (indigenous) people is 1.6 times higher than for non-indigenous people. Among the highest infection rates were aged care residents (of privately owned facilities), people with disabilities, temporary migrants, and multicultural communities.

Once Covid restrictions were lifted and people went back to work, the temporary monetary stimulus flowed through the economy to expenditure. Production worldwide (the supply side) was unable to meet the rise in demand – not only as an aftermath of Covid but also because of physical supply disruptions caused by the war in Ukraine. The result was surging inflation.

The situation in Australia now is not a wage-price spiral, but a price spiral with wages unable to catch up. According to Australian economist Dr Andrew Leigh, now Minister for Competition, Charities, and Treasury in the Albanese Government, what we have is 'a growing body of evidence that suggests excessive market concentration'.[70] The gap between firms' cost of production and their selling prices is a strong indicator of market power (oligopoly). Even before the post-Covid inflation, Australian research led by Treasury's Jonathan Hambur found that industry average mark-ups increased by 6 per cent between 2003 and 2016. 'This fits with figures for the advanced economies estimated in a study by the IMF over the same period'.[71]

It also fits with ballooning profits of large companies. Research by Dr Jim Stanford of *The Australia Institute* found that excess corporate profits account for 69 per cent of additional inflation beyond the Reserve Bank's target, whereas rising labour costs per unit of production, adjusting for the productivity of labour, account for just 18 per cent.[72]

The combined impact of oligopoly and rising interest rates affects the population very unequally. Increased interest rates specifically target consumers in the housing market: middle and low-income earners, those who have recently obtained a mortgage, those seeking to buy a home, and those seeking rental accommodation. As economist Gareth Hutchins points out: 'By lifting rates, it's forcing households with mortgages to hand over more of their money to their banks, via higher interest payments, so those households have less money to spend at the shops so inflation will hopefully fall'.[73]

Neoliberal Regression Thus Far

In conclusion, the neoliberal regression has failed society in many ways. Under neoliberal regimes, inequality has increased everywhere. There is vastly unequal access to the most basic necessities of a civilised life: food, health, shelter, and education. Poverty remains entrenched even in rich countries. The uber-rich hide their wealth in tax havens to avoid paying for what governments provide for society. Instead of investing productively, the rich waste their social licence in pursuit of positional goods such as enormous yachts, palatial houses, digital currency, and non-fungible tokens.[74] The continuing growth of the monetary value of such goods depends essentially on wealth continually being generated and channelled into them. In that respect, items such as Bitcoin are like giant Ponzi schemes dependent on growing inequality.

Neoliberal deregulation has led to global economic crises, undermining the very system that gave the wealthiest people their wealth. In the words of Oxfam's inequality research director Anthony Kamande,

> People are losing jobs in droves, wages are on the decline, the quality of jobs is deteriorating and governments, especially in poorer countries, are facing a fiscal squeeze, all of which are threatening the lives and livelihoods of the world's poorest people.[75]

Neoliberalism has, paradoxically, concentrated power in the hands of political leaders, in an exact reversal of what its Montpèlerin founders wanted, while at the same time diminishing the sources of independent advice from the state bureaucracy. Neoliberal failure to improve the prospects of majorities has opened the way for neo-fascists like Donald Trump and Jair Bolsonaro. Israel now stands on the threshold of autocracy.

The neoliberal regression has undermined the capacity of human society to overcome crises of natural origin. The world failed the (social) distributional test of the Covid pandemic – a social failure masked by the technological success of MRNA vaccine production. With by far the greatest natural threat ever faced by humanity, global heating, societies can no longer be governed by the neoliberal ideological model if they are to survive.

The way forward to a new social transformation is far from clear. There is no single, simple working model of socialism, such as Leninist communism, to turn to. A working, multifaceted alternative has to be invented from the many threads of democratic socialism in the twentieth century and modified for the present. These matters are discussed in the next chapter. What form will the ideological transformation necessary to confront climate change take? How will it be accomplished?

Notes

1 Piketty, T. (2020) p. 7.
2 Christensen, M-B et al. (2023).
3 Piketty T. (2014) Figure 9.5, p. 319.
4 Ibid., p. 349 Figure 10.6.
5 Ibid., p. 244.
6 Ibid., p. 244.
7 Piketty, (2014) p. 71
8 Ibid., Chapter 8.
9 Ibid., p. 303.
10 Ibid., p. 334-335
11 Ibid., p. 116.
12 Russel, K and Smith T.J. (2023) p. 29.
13 Piketty T. (2014) pp. 260-261.
14 Ibid., p. 262.
15 Banerjee, A. and Duflo, E. (2022).
16 Ibid.
17 'The data and analysis presented here are based on the work of more than 100 researchers over four years, located on all continents, contributing to the World Inequality Database (WID.world), maintained by the World Inequality Lab. This vast network collaborates with statistical institutions, tax authorities, universities and international organizations, to harmonize, analyze and disseminate comparable international inequality data'. in Chancel L., Piketty T., Saez E., and Zucman G. (2022) p. 10, 'Executive Summary'.
18 Ibid., p. 11.
19 Ibid, p. 11
20 World Inequality Report, Figure 10, p. 16.
21 Ibid., p. 15.
22 Ibid., p. 15
23 Ibid., p. 15
24 Ibid., p. 16
25 Joseph Rowntree Foundation (2023) p. 3.
26 https://unstats.un.org/sdgs/report/2022/goal-01/ (accessed 29/03/2023). The decline in poverty may well be accounted for by the rise of China.
27 Christensen, M-B et al. (2023) p. 2.
28 Alesina, A., and Glaeser, E.L. (2004).
29 Barber, W. (2003) 'The fourth leading cause of death in the US? Cumulative poverty.' *The Guardian* on line 22/06/2023 (see Brady, Kohler and Zheng, H., 2023, pp. 618–619.

30 https://povertyandinequality.acoss.org.au/poverty/poverty-rates-in-oecd-countries/ (accessed 29/03/2023).
31 Chancel, L., Bothe, P. and Voituriez, T. (2023).
32 IPCC (2023) *AR6 Synthesis Report*, https://www.ipcc.ch/report/ar6/syr/resources/spm -headline-statements/ (accessed 22/03/2023). Section A1.
33 IPCC (2023) Section A1.
34 Ibid., section C1.
35 Chancel, L., Bothe, P. and Voituriez, T. (2023) 'Key Facts'.
36 UN Water (2021) *Water Scarcity*, https://www.unwater.org/water-facts/water-scarcity (accessed 22/03/2023).
37 Chancel L., Piketty T., Saez E., and Zucman G. (2022) *World Inequality Report 2022*, p. 116.
38 Ibid. Figure 6.2, p. 117.
39 https://press.un.org/en/2023/sgsm21730.doc.htm (accessed 26/03/2023).
40 Chancel L., Piketty T., Saez E., and Zucman G. (2022) pp. 117–118.
41 Ibid., p. 118. 'To obtain these numbers, we simply divide the remaining carbon emissions budget by the cumulative global population that will be emitting it over the coming decades. According to the United Nations, there will be 265 billion individual-years between now and 2050. This implies a sustainable per capita budget, compatible with the +2°C temperature limit, of 3.4 tonnes per person per annum between now and 2050. This value is about half of the current global average. The per capita sustainable budget compatible with the 1.5°C limit is 1.1 tonne of CO2 per annum per person, i.e. about six times less than the current global average.'
42 Ibid., p. 118.
43 *World Inequality Report*, 2022, p. 11
44 https://www.ga.gov.au/scientific-topics/national-location-information/dimensions/ australias-size-compared#:~:text=Australia%20is%20the%20planet's%20sixth,is%20 the%20world's%20largest%20island. (accessed 22/03/2023).
45 https://www.abs.gov.au/articles/services-australian-economy (accessed 30/05/2023).
46 https://www.careeconomycrc.com.au/about (accessed 30/05/2023).
47 There is a small fine for failing to register a vote at polling stations or by post. There is no penalty for voting 'informally' – spoiling the ballot paper or otherwise not registering a valid vote.
48 Tingle, L. (2016).
49 Philips, J. (2022). See also: https://povertyandinequality.acoss.org.au/poverty/#:~ :text=Our%202022%20Poverty%20in%20Australia,a%20couple%20with%202 %20children.
50 Ibid.
51 Markham, F., and Biddle, N. (2018).
52 OECD (2019), Poverty rate (indicator). doi: 10.1787/0fe1315d-en.
53 https://www.rba.gov.au/about-rba/history/.
54 https://www.rba.gov.au/education/resources/explainers/what-is-monetary-policy.html.
55 https://www.rba.gov.au/about-rba/history/.
56 Ryan S. and Branston, T. eds (2003).
57 The report, prepared by McGuirk Management Consultants for the Governance Institute of Australia, surveyed 1,167 boards from across the public, private, and not-for-profit sectors, and included 226 share market-listed companies. https://www.governanceinstitute.com.au/. See Janda, M. (2023) 'Wage inflation hits double digits … for chief executives, Governance Institute finds', Australian Broadcasting Commission report 14/06/2023.
58 Australia Institute Media Release (2023) 'OECD report shows corporate profits contributed far more to inflation in Australia than wages', 08/06/2023. https://australiainstitute.org.au/post/oecd-report-shows-corporate-profits-contributed-far-more-to -inflation-in-australia-than-wages/ (accessed 10/06/2023).

59 Wright, S. (2023) 'Bumper profits spurred inflation, reports OECD', Melbourne, *The Age,* 08/06/2023.
60 Reported in *The Guardian Australia* (07.06/2023). https://www.theguardian.com/australia-news/2023/jun/07/rba-governor-philip-lowe-australia-interest-rates#:~:text=%E2%80%9CIf%20people%20can%20cut%20back,particularly%20for%20the%20services%20sector (accessed 10/06/2023).
61 Hutchens, G. (2023) 'The Reserve Bank doesn't need an overhaul, Australia needs fairer policies, says former governor', ABC News 30/03/2023: https://www.abc.net.au/news/2023-03-30/bernie-fraser-former-governor-rba-review-neoliberalism-inflation/102131054 (accessed 30/03/2023).
62 Shorten, B. (2022) 'True extent of Robodebt shame revealed' (Department of Social Sevices) https://ministers.dss.gov.au/editorial/9661 (accessed 14/03/2023).
63 https://robodebt.royalcommission.gov.au/ (accessed 10/07/2023).
64 Hutchens, G. (2023) 'Robodebt was a scandal. Should economists bear some responsibility for it too?' https://www.abc.net.au/news/2023-07-09/robodebt-inflation-targeting-economists-and-unemployment/102575336 (accessed 11/07/2023).
65 Aly, W. (2023) 'We treat unemployed as aliens', Melbourne *The Age,* 10/03/2023 p. 24.
66 Tooze, A. (2021) pp. 34–35.
67 https://treasury.gov.au/coronavirus (accessed 16/03/2023).
68 Australian Bureau of Statistics (2023) *COVID-19 Mortality in Australia: Deaths registered until 31 January 2023,* https://www.abs.gov.au/articles/covid-19-mortality-australia-deaths-registered-until-31-january-2023 (accessed 21/03/2023).
69 Bryant, N. (2023) p. 23.
70 Cited by Gittens, R. (2022) 'Weakening competition is adding to our inflation woes'. *The Age,* 17/12/2022 p. 5. Ross Gittens is Economics Editor of the Sydney Morning Herald.
71 Ibid.
72 Gittens, R. (2023) 'Have a moan about inflation, but just don't mention profits', *The Age,* 11/03/2023 p. 3.
73 Hutchens, G. (2023) 'Is there a better way to quell inflation than raising interest rates?' (https://www.abc.net.au/news/2023-02-12/raising-interest-rates-reserve-and-bank-and-inflation-management/101952926, accessed 16/03/2023).
74 Positional goods: goods which are in limited supply and which become more sought after and relatively more expensive as material prosperity increases. The concept was invented by the economist Fred Hirsch (1977) to describe a category of goods that depend for their value on their relative scarcity, status superiority, and perceived power.
75 https://www.aa.com.tr/en/world/250m-more-people-could-face-extreme-poverty-amid-rise-in-inequality-oxfam-official/2868168 (accessed 12/07/2023).

References

Alesina, A. and Glaeser, E.L. (2004) *Fighting Poverty in the US and Europe: A World of Difference,* World Bank, Oxford University Press.
Aly, W. (2023) 'We treat unemployed as aliens', *The Age,* Melbourne, 10/03/2023 p. 24.
Australian Bureau of Statistics. (2023) COVID-19 Mortality In Australia: Deaths Registered Until 31 January 2023. https://www.abs.gov.au/articles/covid-19-mortality-australia-deaths-registered-until-31-january-2023 (accessed 21/03/2023).
Banerjee, A. and Duflo, E. (2022) 'Foreword', in ChancelL., PikettyT., SaezE. and ZucmanG. eds. *World Inequality Report 2022,* p. 3. https://wir2022.wid.world/ (accessed 02/03/2023).
Brady, D., Kohler, U. and Zheng, H. (2023) 'Novel estimates of mortality associated with poverty in the US', *Journal of the American Medical Association Internal Medicine* 183(6): 618–619.

Bryant, N. (2023) 'How did Covid change Australia', *The Age*, Melbourne, 21/03/2023 p. 23 (accessed 27/03/2023).

Chancel L., Piketty T., Saez E. and Zucman G. (2022) *World Inequality Report 2022.* https://wir2022.wid.world/ (accessed 27/03/2023).

Chancel, L., Bothe, P. and Voituriez, T. (2023) *Climate Inequality Report, Fair Taxes for the Global South 2023*, World Inequality Lab Study 2023. https://wid.world/wp-content/uploads/2023/01/CBV2023-ClimateInequalityReport-2.pdf (accessed 10/06/2023).

Christensen, M.-B. et al. (2023) *Survival of the Richest, How We Must Tax the Super-rich to Fight Inequality*, Oxford: Oxfam. http://doi.org/10.21201/2023.621477.

Gittens, R. (2022) 'Weakening competition is adding to our inflation woes', *The Age*, 17/12/2022 p. 5.

Gittens, R. (2023) 'Have a moan about inflation, but just don't mention profits', *The Age*, 11/03/2023 p. 3.

Hirsch, F. (1977). *The Social Limits to Growth*, London: Routledge & Kegan Paul.

Hutchens, G. (2023) 'Is there a better way to quell inflation than raising interest rates?' https://www.abc.net.au/news/2023-02-12/raising-interest-rates-reserve-and-bank-and-inflation-management/101952926 (accessed 16/03/2023).

Hutchens, G. (2023) 'The Reserve Bank doesn't need an overhaul, Australia needs fairer policies, says former governor', *ABC (Australia) News*, 30/03/2023. https://www.abc.net.au/news/2023-03-30/bernie-fraser-former-governor-rba-review-neoliberalism-inflation/102131054 (accessed 30/03/2023).

Joseph Rowntree Foundation (2023) *UK Poverty 2023 The Essential Guide to Understanding Poverty in the UK*. https://www.jrf.org.uk/report/uk-poverty-2023 p. 3 (accessed 06/06/2023).

Markham, F. and Biddle, N. (2018) 'Income, poverty and inequality', *CAEPR 2016 Census paper no.2, Centre for Aboriginal Economic Policy Research*, Canberra: Australian National University. https://openresearch-repository.anu.edu.au/bitstream/1885/145053/1/CAEPR_Census_Paper_2.pdf (accessed 10/06/2023).

OECD. (2019) 'Poverty rate (indicator)'. http://doi.org/10.1787/0fe1315d-en.

Philips, J. (2022) *Poverty in Australia, a National Perspective*, Canberra: Australian Council of Social Service and the University of New South Wales.

Piketty T. (2014) *Capital in the 21st Century*, Cambridge: Bellknap Press of Harvard University.

Piketty, T. (2020) *Capital and Ideology*, Cambridge: Bellknap Press of Harvard University.

Russel, K. and Smith T.J. (2023) 'The greatest wealth transfer in history is here', *New York Times.* https://www.nytimes.com/2023/05/14/business/economy/wealth-generations.html (re-published in *The Age*, Melbourne, 24/05/2023 p. 29).

Ryan, S. and Branston, T. eds. (2003) *The Hawke Government: A Critical Retrospective*, North Melbourne: Pluto Press.

Shorten, B. (2022) 'True extent of Robodebt shame revealed' (Department of Social Services) https://ministers.dss.gov.au/editorial/9661 (accessed 14/03/2023).

Tingle, L. (2016) 'Political amnesia, how we forgot how to govern', *Quarterly Essay 60*, Melbourne: Schwartz Publishing.

Tooze, A. (2021) *Shutdown, How Covid Shook the World's Economy*, Penguin Allen Lane, pp. 34–35.

United Nations, IPCC. (2023) *AR6 Synthesis Report.* https://www.ipcc.ch/report/ar6/syr/resources/spm-headline-statements/ (accessed 22/03/2023).

United Nations Water. (2021) *Water Scarcity.* https://www.unwater.org/water-facts/water-scarcity (accessed 22/03/2023).

10

REMAKING DEMOCRACY FOR A WORLD OF CLIMATE CHANGE

Introduction

The world in 2023 is on a course to climate destruction, an existential threat to humanity. This is no cold war; it is hot, present, and ongoing. There are many imponderable questions surrounding a new social transformation fit for a world in the grip of climate change.

As climate change progresses, the peoples of the globe will experience drastic changes in the environments in which they live and work, changes in some ways as profound as the destruction caused by war. Climate change could lead either to despair or to a new awareness of sociality. War, as we have witnessed during World War II and recently in Ukraine, has generated increased social cohesion and solidarity. Will climate change give rise to a new political ethos as the ground in which a new ideology can take root and thrive?

Transformation requires political actors as well as activists, as we saw in Part 1 of this book, forming a broad-based social movement. Against the doctrine of Marxism, this social movement will need to emerge from across the whole socio-economic spectrum and not from a single class (as in 'the proletariat'). It will include elite professionals, intellectuals, and political leaders as well as grass-roots actors, workers, and social movements. How, where, and when will a transformative movement emerge?

In this and the next chapter, we will examine each of these questions: ideology, institutions, political ethos, and social movements. At this point, there are no simple answers. All we can do is indicate some possibilities. In this chapter, the focus is on social democracy as an ideology and its embodied institutions. The following chapter will focus on the process of transformation.

DOI: 10.4324/9781003382133-12

The neoliberal ideology is decaying. Many economists and sociologists have listed its faults. Among them are Nobel Prize winners economists Paul Krugman and Joseph Stiglitz, sociologists Wolfgang Streeck and David Harvey, and political scientists Richard Cockett and Peter Self.[1] I have discussed these in my book (2020) *Being a Planner in Society: For People, Planet, Place*. But the most detailed analysis of the specific failures of neoliberalism as well as the formulation of an alternative comes from Thomas Piketty and the World Inequality team. As Piketty has argued, and Polanyi demonstrated, ideology is critical to human social life and social transformation.

On this Piketty is explicit. He writes,

> The approach taken in this book – based on ideologies, institutions, and the possibility of alternative pathways – differs from approaches sometimes characterised as 'Marxist', according to which the state of the economic forces and relations of production determines a society's ideological superstructures in an almost mechanical fashion. In contrast, I insist that the realm of ideas, the political-ideological sphere, is truly autonomous.[2]

What form, then, will a transformative ideology take?

Social Democracy

Thomas Piketty begins *Capital and Ideology* thus: 'Every human society must justify its inequalities: unless reasons for them are found, the whole political and social edifice stands in danger of collapse'.[3] Ideology, he argues, is an essential ingredient of human society. It is, so to speak, part of human social DNA. Inequality cannot be eliminated, but it can be mitigated, and in seeking a modern mitigation of the growing inequality of market-based societies since the 1980s, Piketty follows the path of Karl Polanyi in analysing the facts of specific societies in specific periods. This he does at great length and in much detail. He writes, 'I have tried to provide a reasoned history of inequality regimes'.[4]

Rather than neoliberalism, the term Piketty uses to describe the inequality regime in the twenty-first century is 'hypercapitalism', a regime that fits the neoliberal regression to the conditions of a twenty-first century digitally connected global economy. Hypercapitalism means not just the acceptance of gross and increasing inequality, but also a vision of poverty as inevitable and functional for society, glorification of extreme wealth, and admiration of the virtuosity of its owners.[5]

In *Capital and Ideology*, Piketty sets out an alternative to hypercapitalism. The detailed outcome of the alternative is worked out in Chapter 17 of his book and elaborated in the *World Inequality Report* 2022 and the *Climate Inequality Report* 2023.

A question to think about in what follows is this: Is the mitigation of inequality that Piketty and the World Inequality Lab propose an 'ideology'? Since Piketty ranks social democracy as one of six ideological categories, it seems that the answer may be 'yes'. Or is it simply the application to social institutions of the human qualities of ethics, imagination, intuition, memory, and reason as argued by the philosopher Ralston Saul?[6] We will return to this question after examining its main elements.

Whatever the answer, there is no doubt that ideologies capture the imagination. A single word was used to describe each of the competing ideologies of the twentieth century: liberalism, democracy, socialism, and communism. Each single word concealed the immense complexity of social reality. One-word ideologies are compelling and inspiring, but do not reflect reality. Piketty is forthright in insisting that the new ideology should be called 'socialism', but, recognising the underlying complexity, he then adds a qualifier: *'participatory* socialism'. In this book, we will use 'democracy' as the key term, qualifying it as *social* democracy.

The reason for putting the emphasis on democracy is that the word 'socialism' is still strongly redolent of Marxist–Leninist ideology because that is what inspired communism. Socialism and communism are still, arguably, associated in the public mind (especially in the Anglosphere) with autocracy and the actual failure of freedom, democracy, and justice.

Democracy, on the other hand, is what people all over the world are fighting for when they don't have it and fighting to protect and improve it when they do. Democracy is a unifying meme, whereas socialism is historically associated with division and conflict. Moreover, a form of social democracy was made real in many countries after the Second World War.

It is difficult for anyone today to imagine any regime other than the hypercapitalist globalisation that rules the world today. Ralston Saul asks, 'how can an individual in a position of power even doubt the current ideology without seeming ridiculous?' He quotes British Prime Minister Tony Blair: 'The determining context of economic policy is the new global market. That imposes huge limitations of a practical nature – quite apart from reasons of principle – on macroeconomic policies'.[7]

Piketty argues that the context of globalisation is not and must not be regarded as *determining* national policies. The idea of the inevitability of globalisation as a determining context is itself a central part of the ideology of hypercapitalism. So, in offering an agenda of policies to reduce inequality, Piketty relies on his analysis throughout *Capital and Ideology* to demonstrate that many of these policies already exist or have existed in the recent past.

'A just society', Piketty writes, 'Is one that allows all of its members access to the widest possible range of fundamental goods. Fundamental goods include education, health, the right to vote, and more generally to participate as fully as possible in the various forms of social, cultural, economic and political life'.[8] To achieve such a society, he argues, requires a socialist agenda.

In view of the largely positive results of democratic socialism and social democracy in the twentieth century, especially in Western Europe, I think that the word "socialism" still deserves to be used in the twenty-first century to evoke that tradition as we seek to move beyond it.[9]

In this chapter, I discuss the interwoven strands of Piketty's agenda under eight headings: transparent information, participation and democracy, progressive taxation and redistribution of wealth, the problem of inflation, educational justice, planning, climate justice, and just borders.

Transparent Information

'Globalisation' is not transparent. Wealth is deliberately hidden. One of the most important findings from Piketty's books is the deficiency of comparative, government-mandated information not only about wealth but also about income and the various means of regulating capitalism. These are matters that he and the World Inequality team have sought to correct, working with information that currently exists.

So, for any party that wants to pursue a social democratic agenda, building that information base must be an early and continuing objective. Ideally, such an information base would have as broad an international setting as possible, but there is no reason why governments in the USA, Europe, Australia, Japan, and other countries in East Asia (perhaps eventually including even China) could not agree on the terms of a public register of private assets.[10]

There have already been steps towards corporate tax transparency. Joseph Stiglitz reports that 'The Global Reporting Initiative has developed a tax standard, including country by country reporting, that is regarded as the gold standard. It is supported by large global corporations and some of the world's biggest investors'.[11]

Participation and Democracy

The doyens of American political pluralism, Robert Dahl and Charles Lindblom, in their later work explored the yawning gulf in democracy between what we expect of governments and what we expect of firms. In their view, these are the two main pillars of social power in liberal capitalism. Dahl observes:

> Whatever may have been the reality in the nineteenth century, the twentieth saw the emergence of giant corporations whose governments in both their internal and their external relations took on many of the characteristics of the governments of states. The giant corporation thus became, de facto, if not de jure, both a public enterprise and a political system.[12]

Dahl proposed 'a democratically controlled system of decentralized economic power, a system of self-governing enterprises'.[13] Piketty, therefore, joins a tradition of American political pluralism in proposing that a democracy requires 'sharing power' over decision-making in firms. He discusses the successes and limitations of 'co-management' (existing to this day) in Germany and the Nordic countries. He argues, 'Reducing wealth inequality and capping large shareholder voting rights are the two most natural ways of extending the Germano-Nordic co-management model'.[14]

Perhaps, rather than 'socialism' we might think of Piketty's agenda as 'social pluralism'. Piketty does not entirely rule out state ownership of productive services (electricity, gas, transport, water) where appropriate, but the main thrust of his agenda is consistent with equal, pluralist freedom of the individual.

Piketty goes further than Dahl and Lindblom in re-examining the limitations of the parliamentary model and, 'its inability to stem the tide of rising inequality'.[15] To deal with the imbalance of political power in favour of the wealthy, in the 'financing of political campaigns and of political life more generally', Piketty writes,

> In theory, universal suffrage is based on a simple principle: one woman (or man), one vote. In practice, financial and economic interests can exert an outsized influence on the political process, either directly by financing parties and campaigns or indirectly through the media, think tanks, or universities[16]

Piketty proposes completely prohibiting 'political donations by firms and other moral persons' and instead instituting a system of 'democratic equality vouchers' – one voucher of specified monetary value to be issued to each person of voting age.[17] The conditions placed around the institution of democracy vouchers are discussed in detail by Piketty himself and by Julia Cagé.[18]

Progressive Taxation and Redistribution of Wealth

Piketty's historical investigation leads him to the conclusion that excessive concentration of wealth in the late nineteenth and early twentieth centuries exacerbated social tensions, led to violent nationalism and blocked the social and educational investments that made the balanced post-war development and economic growth possible. Since 1980, he argues, the capitalist world has drifted back towards extreme wealth concentration, and that process continues (as we saw in the previous chapter).

Existing estimates suggest that about 36 per cent of multinational profits are transferred to tax havens each year. In June 2021, more than 130 countries and jurisdictions agreed that multinational profits would be subject to a minimum tax of 15 per cent. Table 10.1 shows the revenue that could have been derived in 2021 from a universal corporate tax of either 15 per cent or 21 per cent.

In order to redistribute wealth and provide for the fundamental goods of a just society and climate justice,[19] Piketty proposes major reform of fiscal regimes. A direct redistribution of wealth from the top 10 per cent to the rest of the population would be achieved by a progressive wealth tax which would fund a 'universal

TABLE 10.1 Revenues from corporate tax of a global minimum of 15 per cent and 25 per cent, 2021

Country	Revenue gained from min. 15% tax (US$ billion)		Revenue gained from min. 25% tax (US$ billion)	
	Without carve-out	With carve-out	Without carve-out	With carve-out
ANGLO				
USA	40.7	38.8	165.4	142.5
Canada	16.0	11.6	34.7	24.9
Britain (UK)	unavailable			
Australia	2.3	1.9	11.7	8.9
EUR/JP				
France	4.3	3.8	26.1	20.7
Sweden	1.5	1.3	5.3	4.2
Germany	5.7	4.8	29.1	22.0
Japan	6.0	5.1	28.7	23.2
EMERGING				
Russia	unavailable			
India	0.5	0.4	1.4	1.2
Indonesia	unavailable			
China	4.5	3.5	30.2	13.1
South Africa	0.6	0.4	3.0	2.3
Argentina				
Chile	0.2	0.1	1.2	0.7
Brazil	0.9	0.8	7.4	5.7

Source: Figure 8.1 World Inequality Report 2023, p. 152.

capital endowment' to be conferred on each adult. Carbon taxes, preferably of individual emissions, would pay for the costs of global heating: pre-emptive, adaptive, and restorative as discussed in Chapter 7 of this book.

Historically, progressive taxes on inheritance and income in the USA and Britain 'ran as high as 70–90 per cent on the highest incomes and largest fortunes for decades – decades in which growth rose to unprecedented levels'.[20] Economists Joshua Gans and Andrew Leigh agree that

> In the post-war decades, macroeconomic growth was strong, and equally distributed. In fact, from 1946 to 1980, incomes for the bottom half of the US population grew faster than incomes for the top half. But this came to an end in the 1980s.[21]

Piketty points out, however, that even with the post-war income and inheritance tax regime, wealth nevertheless remained extremely concentrated. The diffusion of wealth never really touched the bottom 50 per cent. Since the 1980s, the share of private wealth held by the disadvantaged (50%) and even the middle classes (the fiftieth to ninetieth percentiles) 'has shrunk nearly everywhere'.[22] Net wealth includes financial and nonfinancial assets net of debts, including bonds and mutual funds, pensions, housing, public equity, and private business assets, but excluding human capital, durable goods, non-profits, and unfunded defined benefit pension plans.[23]

The concentration of wealth at the top greatly limits the opportunity of the bottom 50 per cent to participate in economic life: 'This is not the ideal of participation that a just society should strive to achieve'.[24] Lack of economic participation may be linked to 'political secession' discussed in Chapter 8.

Piketty's solution is a universal capital endowment to be conferred on every person when they reach the age of, say, 25.[25] Piketty finds a kind of precedent in agrarian reform: breaking up and redistribution of ownership of large landed estates. He cites successful examples: Ireland and Spain in the late nineteenth and early twentieth centuries, Mexico after the 1910 revolution, Japan and Korea after World War II, and Indian States (Kerala and West Bengal) in the 1970s and 1980s.[26]

Piketty argues that extreme inequality of wealth tends to reproduce itself in many ways, so redistributing wealth in all its forms (and not just landed capital) is logical and necessary. Therefore, a progressive property tax would fund the universal capital endowment.

Wealth taxes have been levied for many years in OECD countries: property and land taxes such as the Council tax in the UK, and inheritance tax in the USA and UK.[27] These taxes are simply proportional to the amount of wealth, and therefore not progressive. Norway, Belgium, France, Spain, Italy, and Switzerland already levy wealth taxes either on all or on selected assets.[28] Wealth taxes on individuals – including property and inheritance taxes – typically generate 2–3 per cent of national income in rich countries, 1 per cent in middle-income countries, and 0.5 per cent in low-income and emerging countries[29].

Existing wealth taxes suffer from various deficiencies: competition among countries to keep taxes low, low exemption thresholds meaning heavy impacts on smaller wealth holders, the existence of tax havens, self-assessments rather than systematic information reporting, and ideological and political opposition to wealth taxation. All of these hurdles can be overcome, and there has already been some progress internationally on most of them.[30]

The authors of the *World Inequality Report* (2022) argue that

> Given the enormous increase in the aggregate value and concentration of private wealth in recent decades, it would be completely unreasonable not to ask more of top wealth holders in the future, especially in light of the social, developmental, and environmental challenges ahead.[31]

The *World Inequality Report* proposes a progressive tax on wealth under three scenarios (Table 10.2). The tax rates are marginal rates; they apply to the

TABLE 10.2 Property tax and income tax rates

Progressive property tax (financing the capital endowment for each young adult)			Progressive income tax (financing the basic income scheme and the social and ecological state)	
Multiple of average wealth	Annual property tax (effective rate)	Inheritance tax (effective rate)	Multiple of average income	Effective tax rate (including social taxes and carbon tax)
0.5	0.1%	5%	0.5	10%
2.0	1.0%	20%	2	40%
5.0	2.0%	50%	5	50%
10	5.0%	60%	10	60%
100	10%	70%	100	70%
1,000	60%	80%	1,000	80%
10,000	90%	90%	10,000	90%

Source: Table 17.1 Piketty, T. (2020) *Capital and Ideology,* p. 982.

fraction of wealth possessed between thresholds and not to total wealth. The Report states that

> Rates reach 50 per cent over $10 billion and 90 per cent over $100 billion. This would in effect ban wealth accumulation over $10 billion. The revenue generated in this scenario is equivalent to 5.3 per cent of global income after tax evasion.

But also, 'Such a wealth tax scenario cannot raise such revenue for ever as it effectively prevents decabillionaires and especially centibillionaires from keeping their wealth'.[32]

The World Inequality Report contains a work-out of the funds (as a percentage of global income) that would be available to governments under the three wealth tax scenarios for eight world regions: Sub-Saharan Africa, South and South-East Asia, Latin America, Middle East and North Africa (MENA), East Asia, Europe, Russia and Central Asia, and North America (from Figures 7.3 a–h in the *World Inequality Report,* pp. 141–143)

Global income is the sum of all national incomes (not just the sum of nations included in the seven regions). National income is the sum of all incomes received by individuals in a given country over a year. In 2021, global income amounted to €86 trillion (US$122 trillion), while global net wealth amounted to six times this value, €510 trillion[33].

Recipients of the universal capital endowment would be free to use it as they wished:

> With the system proposed here, every young adult could begin his or her personal and professional life with a fortune equal to 60 per cent of the national average, which would open up new possibilities such as purchasing a house or starting a business.[34]

TABLE 10.3 Revenue from billionaires

Wealth owned US dollars	Number of Adults	Total wealth US$ billion	Revenue % global income Tax scenario 1	Revenue % of global income Tax scenario 2	Revenue % of global income Tax scenario 3
Sub-Saharan Africa					
1 billion to 10 billion	11	52	0.1	0.1	0.3
South and SE Asia					
1 billion to 100 billion	260	991	0.3	0.6	2.5
Latin America					
1 billion to 100 billion	105	419	0.2	0.4	1.5
MENA					
1 billion to 100 billion	75	182	0.1	0.1	0.5
East Asia					
1 billion to100 billion	838	3,446	0.2	0.7	3.6
Europe					
1 billion to 100 billion	499	2,364	0.33	0.6	2.5
Russia & Central Asia					
1 billion to 100 billion	133	586	0.5	1.2	4.1
North America					
1 billion to 100 billion	835	4,822	0.5	1.0	5.0
TOTALS	2,756	12,862	2.23	4.7	20.0

Source: Figure 7.1 and 7.3*a–h* World Inequality Report, pp. 141–143.

Note: Tax scenarios: effective progressive tax rates: 4.5 per cent for US$1–10 billion, 6.4 per cent tax for US$10–100 billion, 8.3 per cent tax for over US$100 billion..

The capital endowment would redistribute wealth from older to younger age groups, providing greater certainty about when children will gain capital (parental inheritance usually coming uncertainly and late in life). However, unless housing supply increases proportionately, and if many young people use the capital to purchase a house, the price of housing may then rise, benefitting existing property owners. So, we need to consider the matter of inflation more generally.

Inflation

Inflation is a process that destabilises the value of money, making exchange through the medium of money difficult, especially when national currencies are used to buy goods on the international market. Today (2023), in the OECD, the danger of inflation looms large. Supply of essentials has been unable to match post-Covid demand. It has also become clear that the neoliberal method of curbing inflation has unjust consequences (see Chapter 9).

What causes inflation? Superficially, inflation is caused by an amount of demand chasing an amount of supply of goods and services insufficient to meet the demand. The contrary is deflation. Yet one of the world's leading economists, Olivier Blanchard, argues that inflation has a deeper cause. Inflation is fundamentally the outcome of a distributional conflict between firms, workers, and taxpayers. Blanchard argues that an inflationary spiral, 'could start from a desire by workers to increase their real wages, or from firms to increase their profit margins, or from the attempts by both sides to maintain the same wage and price in the face of an adverse supply shock'.[35]

Why is inflation a problem? That depends on what the demand is for. If it is for essentials such as housing, medical treatment, food, education, energy, or perhaps raw materials for industry, then it is a problem. If it is for non-essentials, say luxury yachts, brand handbags, or watches, then the market, responding to the price increase, should lead in time to more luxury yachts, handbags, or watches being produced. So, that is not a problem of social justice. As the market adjusts, workers in luxury industries should be able to find employment elsewhere with financial support during short periods of unemployment.

Of particular importance is the effect of inflation on the climate crisis. As governments enforce laws to reduce carbon emissions, they *intend* production systems to respond by introducing low or zero-carbon techniques. However, timing is critical. If there is a delay in the production response, the result will be price inflation.

In neo-classical economic theory, time has no place. In a free market, demand and supply are just supposed to balance in the long run. In the real world, time is critical. Whole industries cannot shift quickly, and there are reasons why they may not want to. Time lags in the response of production to demand can cause inflation.

We have seen this happen to energy production. The demand for energy is not matched by low-carbon production of energy. The result is soaring prices for fossil fuels such as coal and oil. In recent years, this effect has been magnified by supply shortages caused by the war in Ukraine and the oligarchic concentration of capital in a few huge companies. In these circumstances, the price incentive, instead of working to reduce carbon emissions, starts working to stimulate investment in fossil fuel mining and production: a perverse result for the climate crisis.

Any major redistribution in income or wealth can have potentially inflationary consequences, take, for instance, housing. If redistribution accelerates the demand for housing, as it should, unless the supply of housing is simultaneously increased, the price of housing will rise. The result will be a transfer of money to the banks, via mortgages, and existing home-owners. The same tendency will occur for all essential products and services.

Since the 1980s, the problem of inflation has been seen as a wage-price spiral (rather than a price-wage spiral), with pressure from powerful trade unions achieving higher wages, resulting in firms increasing prices for goods and services. The solution is higher productivity: traditionally measured in terms of *more* goods and services for the same or reduced input of labour and resources. But why should we not measure increased productivity as *better* goods and services? Productivity measured against value rather than quantity: better, fairer education and health services, for example. Higher progressive taxation could help suppress inflation and also fund a rapid increase in both the supply and quality of essentials such as housing, health care, and education.

The single instrument for managing inflation/deflation under the neoliberal regression has been the manipulation of interest rates by central banks freed from control by governments. Raising interest rates targets both businesses paying off loans and homeowners with mortgages. Paradoxically, those additional costs add to inflation.[36]

Despite the rhetoric of productivity, the neoliberal solution has been to reduce demand via harsh restrictions on the unions, abolishing the goal of full employment, demonising the unemployed, and increasing the fear of unemployment by reducing the real value of unemployment relief.

In the post-war period when reserve banks were charged with sustaining full employment, a certain level of unemployment at around 2 per cent was recognised as inevitable as workers changed jobs, with short delays in finding new ones. Since the 1980s, unemployment has been explicitly used as a tool to contain inflation. The 'sustainable' rate of inflation was redefined as the 'non-accelerating inflation rate of unemployment' of 3–6 per cent (noted in the previous chapter).

Both Piketty and the authors of the World Inequality Report spend little time addressing solutions to inflation. Yet Piketty recognises the problem and hints at solutions:

> Let us recall that the only true limit to monetary policy is inflation. So long as there is no substantial increase in consumer prices, there is no solid reason not to increase the money supply if it enables us to finance useful policies such as the struggle for full employment, a guaranteed job, the thermal insulation of buildings, or public investments in health care, education, and renewable energy. Inversely, if inflation flares up in the long term, then that means the limits of monetary creation have been reached and that it is time to rely on other tools to mobilise resources (beginning with taxes).[37]

Inflation caused by insufficient supply to meet surging demand hit home in 2023. The application of the singular tool of the reserve banks has created gross social injustice and threatens economic recession in countries where it has been savagely applied – especially in the USA.

It is beyond the scope of this book to investigate in any depth alternatives to the neoliberal solution to inflation. But alternatives exist and are increasingly being discussed by independent economists, the IMF, and even officials of reserve banks. The following are some of the matters under discussion.

- Concentration of capital into bigger and bigger conglomerates is a feature of economies under the neoliberal regression. Inflation of profits and prices can be caused by insufficient competition among big companies if there are too few of them operating in certain sectors. Therefore, stricter and more effective regulation of company takeovers is necessary.
- Fiscal policy needs to be coordinated with monetary policy. It makes no sense for governments to lower taxes and increase special grants (for instance, for house purchase) while central banks raise interest rates. Fiscal policy needs to be directed at increased supply (for instance, of housing).
- The need for complete independence of reserve banks to determine the desirable level of inflation and unemployment is being questioned. Such matters are profoundly political, and the parameters should be decided by democracy, even if independent reserve banks remain in control of interest rates within those parameters.
- Taxes reduce spending power for the sections of the population that are taxed. Wealth taxes on the top 50 per cent would reduce spending on non-essentials and luxuries. But increasing spending power for the bottom 50 per cent would still, in the short run, lead to inflation in the price of essential goods they want to buy. While production of such goods catches up, inflation can be contained by enforced saving, for instance, by paying a proportion of wages into pension funds or superannuation accounts to be accessed on retirement.
- Alternatively, in the shorter run, instead of wage earners paying more income earned to the banks in higher interest rates, a proportion of income might be temporarily placed in holding funds to be released after a surge in inflation is contained. This solution was first canvassed by Maynard Keynes in his 1940 essay *How to Pay for the War*.

Economists talk about inflationary expectations as though they were a personal matter. Perhaps the worst of inflation is caused by such expectations. But these 'expectations' are not just individual judgements, but endemic in the political ethos of a national society. Argentina, for instance, was in 2023 experiencing hyperinflation of more than 100% per cent per year. Argentina has an independent central bank. The bank interest rate was 78 per cent.[38] Under these circumstances, goods have more value than pesos. So those who have money buy up goods today because they will cost more tomorrow.

Educational Justice

One of the most striking charts appears early in *Capital and Ideology* (Figure 1.8, p. 35). It shows the curve of the relationship in the USA between parental income and university access. It is a straight line stretching from barely 30 per cent (university access) for children of the poorest 10 per cent to 90 per cent access for children of the richest 10 per cent. Piketty argues, 'History shows that economic development and human progress depend on education and not on the sacralization of inequality and property'.[39] So educational justice is correlated with economic development.

Piketty acknowledges the practical complexity of finding a fair balance in educational funding. He asks, 'How should one think about a just distribution of public educational investment in a country like France?'[40] He answers: 'A relatively natural norm would be that every child should have the right to the same educational funding, which could be used for either schooling or other training'. He also notes that 'All studies show that early intervention, particularly in primary and middle school, is the best way to correct scholastic inequality between students of different social backgrounds'.[41]

In this respect, it is shocking to learn of the prevalence of illiteracy in a rich country, Australia. People with poor reading and writing skills are at risk of not being able to participate fully in the labour market, education, and civic life. According to the Australian Bureau of Statistics, around 3.7 per cent of Australians aged 15–74 were functionally illiterate, unable to read even road signage or instructions. A further 10 per cent were at the lowest literacy (Level 1). More than 40 per cent lacked literacy to read a book or get an education.[42]

In an Australian study of inequality of education, Associate Professor Laura Perry argues that educational outcomes result both from the home (and community) situation of the child and the quality of the school. Social disadvantage at home reduces parent and care-givers' capacity to support children's school learning, but by Year 3 the influence of the school is just as strong.[43]

One way to reduce educational inequality is to reduce parental poverty. But important also is socio-economic segregation amongst schools. 'Segregated schooling, which occurs when socially advantaged students are segregated into some schools and socially disadvantaged students are segregated into other schools, is neither efficient nor effective'.[44] The OECD Program for International Student Assessment (PISA) measures segregation (among many other factors). Canada has one of the least segregated schooling systems in the OECD, and Australia has one of the highest. Advantaged students have the same performance on PISA in the two countries, but low SES students perform substantially better in Canada than in Australia.

Education is not only associated with human and economic development but also an essential support for democracy, as liberal philosopher John Stuart Mill insisted. The ability to assess the truthfulness of information (for instance coming through the social media) is today a critical human capacity, the absence of which has important political as well as personal consequences.

Toward an Ecological Welfare State

Piketty defines rising inequality and global heating as the greatest challenges facing societies today. A tax on carbon emissions is unquestionably the most effective way to mitigate global heating, but it must be accompanied by compensation for those adversely affected. 'At a minimum, all proceeds of the carbon tax must be put toward financing the ecological transition, particularly by compensating the hardest-hit low income families'.[45]

Piketty goes further, however. He points out that a small group of consumers of carbon-intensive goods and services is responsible for a large proportion of global carbon emissions: 'individuals with high incomes and large fortunes living in the world's wealthiest countries.'[46] Further, the lifestyle changes necessary for the world to manage the climate crisis are so great that these changes are unlikely to be politically acceptable 'without establishing stringent and verifiable norms of justice'.[47]

Here, then, Piketty opens up a question which has been wilfully ignored in the mainstream of political debate. Can global heating be mitigated without far-reaching lifestyle change? We will probably know the answer within the next 20 years. If Piketty is right, then his proposition that a progressive carbon tax to be applied at the level of individual consumers may well become realistic.

Meanwhile, the *Climate Inequality Report 2023* reminds us that the Paris Climate Agreement is 'a legally binding international treaty on climate change'.[48] The Agreement binds all 196 nations and requires economic and social transformation. Article 2.1 states that in order to hold climate change well below two degrees above pre-industrial levels, it would mean 'making finance flows consistent with a pathway towards low greenhouse gas emissions and climate-resilient development'. Article 9 states that developed countries 'should continue to take the lead in mobilizing climate finance from a wide variety of sources, instruments and channels, noting the significant role of public funds, through a variety of actions, including supporting country-driven strategies.'[49]

In accordance with Article 2.9, the accompanying COP21 decision states that 'developed countries intend to continue their existing collective mobilization goal through 2025 in the context of meaningful mitigation actions and transparency on implementation'. Before the 2025 Conference of the Parties, a new collective quantified goal shall be set 'from a floor of US$100 billion per year, taking into account the needs and priorities of developing countries.'[50]

The *Climate Inequality Report* stresses 'the need to further mainstream distributional analysis in climate adaptation and mitigation policies, and finance programs around the world (in low, middle and high-income countries)'.[51] The Report states that 'including investment and spending for mitigation, adaptation, and loss and damage that will amount to $6,300 billion worldwide in 2030 and should be about $4,200bn in 2021'.[52]

This figure may well be an underestimate. A study of Australia's transition to net zero emissions by the 'Net Zero Project' reports that trillions of dollars (AU$) will need to be spent to decarbonise Australia's economy.[53] The study reports joint research by Princeton University in the USA and the Australian Universities of Melbourne and Queensland. The best contribution Australia can make to global decarbonisation is to stop exporting fossil fuels.

The *Climate Inequality Report* addresses the costs of climate change and their distribution in terms similar to the categories mentioned in Chapter 7 of this book, though in somewhat different language.

Preemptive Costs

Pre-emptive action to reduce global heating, the Report argues, 'remains the best recipe for tackling climate inequality'. However, such action, if it is to be effective, incurs adaptive costs: 'accelerating mitigation programs may disproportionately increase economic stress on certain segments of the population within countries. In such cases, it will be crucial to offer generous support mechanisms to vulnerable actors (whether households or firms).'[54]

Adaptive Costs

Adaptation, the Report argues, remains 'vastly underfunded'. Adaptive costs are incurred in rich countries when industries change, and workers need retraining if they are not to be made unemployed. Housing will need to be refitted with effective insulation and cooling. Cities will need to be redesigned to reduce the heat island effect. Transport systems will need to be replanned for low carbon impact, including provision for walking and cycling. The means of electricity supply will have to be rethought and replanned. All of these costs are multiplied in developing countries, some of which do not have even basic infrastructure.

The Climate Inequality Report states that 'the costs of adapting to climate change in developing countries are substantial and rich countries have committed to scale up support for adaptation in developing countries'.[55] Promises were made to double adaptation finance between 2014 and 2020 under a road-map presented to COP 22 in 2016 – and reiterated between 2019 and 2025 following COP 26.

However, where international funding falls short, progressive wealth taxes on the richest wealth holders worldwide (the top 0.1%) could generate substantial funds without asking more from 99.9 per cent of the population.

Individual-based levies such as air passenger taxes and progressive wealth taxes, or taxes on specific, polluting economic sectors of the economy can also be mobilized. The removal of fossil fuel subsidies can also save significant amounts of funding, but careful design and timing are critical.[56]

Restorative Costs (Loss and Damage)

Beyond adaptive costs, populations vulnerable to weather effects will require restitution when subject to floods, storms, and droughts. This category of costs is termed by the UN 'loss and damage'. The decision made at COP 27 to create a 'Loss and Damage Fund' is a step in the right direction. Yet the timeline for operationalising the fund is quite short, by COP 28 [November 2023], and politically sensitive questions remain on who benefits and who pays'.[57]

The Climate Inequality Report shows that a global progressive wealth tax on centimillionaires (with net wealth between 100 million and 1 billion US dollars) taxed at 1.5 per cent would raise $US109 billion annually. A tax of 2 per cent on wealth between 1 billion and 100 billion dollars and 3 per cent on wealth over 100 billion would yield a further 187 billion dollars.[58] This contribution would go some way to meeting the costs of compensation for loss and damage to populations suffering from global heating that they had no part in causing. But it will not be enough.

In the longer term, the climate finance agenda is necessarily dependent on a taxation strategy to build an 'ecological welfare state' for all nations during the twenty-first century which would ensure a just transition in line with the objectives of the Paris Agreement.[59]

The Climate Inequality Report examines overall needs from existing international funding sources. The total climate finance in 2021 amounted to US$850 billion compared with the needed funds estimated by Naran et al.[60] A high-level expert report from the London School of Economics found that part of the total investment and spending needs in developing countries is already covered by existing and planned investments. Hence, the additional investments required would be about $1,800 billion per year in 2030 for low- and middle-income countries excluding China.[61]

The LSE report states: 'Without appropriate adaptation action to ensure that the most vulnerable populations have the financial, technical and institutional resources to cope and recover from climate-related events, climate change could push more than 100 million people below the poverty line by 2030'.[62]

Strengthening the Nation-State

In order to make a social transformation adequate to meet the climate challenge, governments need the capacity to plan. To win World War II required military and civil planning on an unprecedented scale. Everyone knew that food supplies had to be planned and distributed fairly. The vision was of equal opportunity to prosper for all. Housing had to be provided to replace the dwellings destroyed, and more than that: to build a future living environment better than the past. New towns and old had to be planned. Agriculture and the countryside had to be protected. The town planning profession was reformed to create new and powerful institutions of planning and development control.

Historian Tony Judt, whose work we met in Chapters 3 and 4 put the matter very clearly:

> On one thing, however, all were agreed – resistors and politicians alike: 'planning'. The disasters of the inter-war decades – the missed opportunities after 1918, the great depression that followed the stock market crash of 1929, the waste of unemployment, the inequalities, injustices and inefficiencies of laissez-faire capitalism that had led so many into authoritarian temptation, the brazen indifference of an arrogant ruling elite and the incompetence of an inadequate political class – all seemed connected by the utter failure to organize society better. If democracy was to work, if it was to recover its appeal, it would have to be *planned*.[63]

Today, since the neoliberal regression, the capacity of governments to plan has been all but destroyed. Planning requires a strong professional public service with the capacity to give independent advice to the political tier and command over funds. Individuals have become richer, some with enormous wealth, but governments have become poorer. This trend was magnified by the Covid crisis during which governments borrowed the equivalent of 10–20 per cent of GDP. As the authors of the *World Inequality Report* state, 'The currently low wealth of governments has important implications for state capacities to tackle inequality in the future, as well as the key challenges of the 21st century such as climate change'.[64]

In many countries, the expenditure of the state, variously called the public service, civil service, or the public sector, has expanded enormously. Yet, the power of the state to meet the needs of the public has been drastically weakened. This tendency resulting from the neoliberal regression has been described by David Harvey in his history of neoliberalism: for example, in Eastern Europe, South Asia, China, and South America.[65] The state has been weakened by selling off public utilities to the private sector in what Harvey describes as 'accumulation by dispossession': the transfer of a public asset which can generate income for a government over the long term into private hands for the purpose of enhancing private profit.[66]

Piketty remarks that from the 1950s to the 1970s, capitalist countries were mixed economies (though they varied in form in different nations). Public assets took many forms, including infrastructure, public buildings, schools, and hospitals; in addition, many firms were publicly owned, and there was public financial participation in certain sectors. Furthermore, public debt was historically low owing to post-war inflation and government measures to reduce debt such as exceptional taxes on private capital or even outright debt cancellation.[67]

Between 1950 and 1980, the share of public capital net of debt was generally around 20–30 per cent of total national capital (25–30% in Germany and Britain, 15–20% in France, the USA, and Japan). These figures gave states considerable power over their national economies. But since the 1980s, Western countries 'have long since ceased to be "mixed economies"'.[68] Privatisation of public assets, along with limited investment in the remaining public sectors especially education and health, and the steady increase in public indebtedness have shrunk public capital in the national economy 'to virtually zero'. In the USA and the UK, it is negative: public debts exceed the value of public assets. The United States national debt exceeded 31 trillion dollars by the end of 2022. It stands now at more than 120 per cent of GDP, greater per GDP than in the post-war years 1945–1946.[69]

The Australian social economist David Hayward in an examination of the Australian State of Victoria writes,

> The state, then, has not only massively expanded, but this has been accompanied by the rise of monopolies, which in turn manage to pay very little tax on their total earnings. Profits have been privatised and risks socialised. There is very little if any evidence to show that 'consumers''have been the main beneficiary or that services have improved[70]

Hayward refers to a recent book by the British political economist Brett Christophers on rentier capitalism.[71] Rents are income derived from the ownership, possession, or control of scarce assets under conditions of limited or no competition. What Christophers means by rentier capitalism is 'an economic system not just dominated by rents and rentiers but, in a much more profound sense, substantially scaffolded by and organized around the assets that generate those rents and sustain those rentiers'.[72]

An important part of this scaffolding is public and private contracting and subcontracting. Contracts are awarded for services to be provided. The contracts themselves become prized assets of companies benefiting from outsourcing, especially of services provided to governments which would formerly have been provided internally by the state. These contracts are limited in number and encompass service delivery over a period of years, sometimes decades. They are scarce assets generating a form of rent 'income guaranteed by virtue of possession of an asset that insulates the contractor from all competition for the contract duration'.[73] The key skill for contractors is winning contracts, the longer the better, 'for once won, the winner is shielded from competition for its duration'.[74]

The problem is that rent is value derived from *possession* of a scarce asset, rather than value derived from *production* of desired goods and services involving investment in innovation leading to higher productivity. The rise of 'rentierism', Christopher argues, is responsible for the feeble growth of productivity in Britain, and we might add, in other Western countries. Rentierism weakens the state by outsourcing what were previously public services provided by the state to firms gathering rent from lucrative contracts in many different fields from public health, aged and disability care, and education to housing and transport infrastructure.

Even more importantly, outsourcing economic advice to private consultancies (as discussed in Chapter 8) means that the interests of global capital and neoliberal orthodoxy have been taken into the heart of what was formerly 'public service'.

Just Borders and Transnational Justice

Developing new forms of democracy, fiscal progressivity, educational, and climate justice may require constitutional reform. The path to such reform varies greatly among democratic nations. In the United Kingdom, the constitution 'is composed of the laws and rules that create the institutions of the state, regulate the relationships between those institutions, or regulate the relationship between the state and the individual. These laws and rules are not codified in a single, written document. In effect, Parliament decides'[75] – as it did when Britain joined the European Union and when it left. The Indian Constitution follows the British model, but constitutional change requires a super-majority of two-thirds of members of Parliament. Amendment of the Australian Constitution is by referendum with simple majority approval in federal Parliament and in a majority of States.

In the USA, constitutional amendment is much more difficult. It must be proposed by two-thirds of both houses of Congress or two-thirds of the States. Ratification requires majority votes in three-quarters of State legislatures. Even so, the United States Constitution has been amended 27 times.[76] The introduction of federal income tax and inheritance tax required ratification of the Sixteenth Amendment to the Constitution in 1913. The latest Amendment was ratified in 1992.

Piketty argues, 'To shrink from changing the rules because it is too complicated is to ignore the lessons of history and forgo any possibility of real change'.[77] New challenges that lie ahead – as for example generated by global heating – may generate irresistible public demands for 'real change'. The nature of political allegiances creating two main parties may even itself change.

Finally, Piketty comes to what he says is 'the most delicate question': the question of just borders and transnational justice. He continues, 'One of the most obvious contradictions of the current system is that free circulation of goods and capital is organized in such a way that it significantly limits the ability of states to choose their fiscal and social policies'.[78] National sovereignty over such policies is 'fictional'. The requirements of justice are 'increasingly transnational'. Gradually, this reality is already being recognised in small ways, for instance, the transnational move to enforce a minimum 15 per cent corporate tax rate designed to reduce the appeal of tax havens.[79] As long as governments do not outsource their research and policy making to global consultancies like PwC, which then leak the proposals to their multinational clients – see Chapter 8.

Piketty proposes a new model of globalisation to include transnational democracy to make decisions regarding global public goods: protecting the environment, promoting research (including into inequality and poverty), and investigating the possibility of imposing common taxes on income and property, on large firms, and on carbon emissions in the interest of global fiscal justice.

In principle, this model is similar to that of David Held who proposed the concept of 'cosmopolitan democracy' to address the erosion of democratic autonomy by glo-balisation. States, he argues, negotiate ad hoc agreements with one another in the presence of a complicated nexus of pre-existing international ties: 'The operation of states in an ever more complex international system both limits their autonomy (in some spheres radically) and impinges increasingly upon their sovereignty'.[80]

A new democratic order is required that is 'cosmopolitan', that is to say, it must extend beyond national boundaries and ultimately involve the whole world. Cosmopolitan democracy does not entail a single global structure of governance, but it does entail a mixture of institutions encompassing different territorial scales, some of which will be global.

Piketty's concept, unlike Held's, does not envisage global democratic insti-tutions such as a global parliament. Instead, Piketty envisages 'transnational assemblies' under treaties negotiated by nations sharing common interests. Such assemblies 'could be composed of members of the national parliaments of member states or of transnational deputies expressly elected to serve in this capacity, or a mixture of the two'.[81]

There are many ways to organise transnational assemblies, Piketty argues, and it is reasonable to experiment with different solutions in different contexts. Mutual interests and solidarities are politically constructed and come into being when people see that 'the advantages of belonging to the same community [of interests] outweigh the advantages of maintaining borders'.[82]

Piketty argues that 'the return of social progressivity and the implementation of a progressive property tax should take place in as broad an international setting as possible'.[83] Such a setting would allow governments to establish public financial registers with all pertinent information about the ultimate owners of wealth in various countries.

That said, however, there is plenty of scope for national governments to make progress towards greater equality and the elimination of poverty. As mentioned above, well within the scope of national governments are: developing precise, consistent national information on who owns what wealth, levying wealth taxes and progressive corporate taxes, mandating worker representation on company boards, eliminating the power of wealth in democratic voting, and finding fairer and more democratic ways of containing inflation.

A false fatalism about globalisation, assuming that one unique policy must be imposed on everyone, is responsible for the abandonment of ambitious economic reforms and the retreat into nativism and nationalism. Piketty points out that large economic powers such as the United States and perhaps China and India already have the means to enforce any decisions they make regarding taxes. US tax law applies to US citizens wherever they live.[84] There is no reason why the same should not apply to other citizens who value the countries in which they live and do not wish to relinquish citizenship of those countries.

Conclusion

Is social democracy an ideology? That must remain a moot point. All of the proposed reforms, Piketty insists, are the start of a new *conversation* for democracy and social justice. The past examined by Polanyi and Piketty is the enduring social history of human societies and their dependence on ideology to explain the continuity of inequality.

Piketty's future vision is a long-distance one as he imagines alternative ways to confront the challenges of social and economic inequality and global heating. New problems of environmental damage will emerge from over-exploitation of Nature in the future as they have in the past, some known, some unexpected. Today, we must include the challenge of viral infection. We have not heard the last of pandemics or of the costs of keeping economies afloat in their presence.

Ideology cannot be transcended by some mythical socialist science containing a unique truth. Ideology must always be contested and open to change. It is the lazy consensus on the idea that the global market will solve all social problems and that current policy constraints are inevitable, that is pernicious for human progress. It is the fearfulness of the political class to open 'Pandora's Box' of imagined evils that so often prevents alternative paths from being explored and social experiments from being undertaken.

In the ancient Greek legend Pandora meant 'all-gifted', among which was the gift of curiosity. She was given a beautiful box by Zeus but was told never to open it. As payback for Prometheus who had created humanity, Zeus had packed the box with all the evils that could afflict humans. Naturally, Pandora, being curious, opened the box. Disease, madness, violence, and death flew out. But Prometheus had smuggled 'hope' into the box. In hope, alternative paths *are* being explored and have been in the past with successful outcomes.

Piketty's political position is, I believe, profoundly pluralist, acknowledging cultural, social, national, and local differences as valuable and essential. His proposed agenda will seem didactic, but its purpose is to get down to the level of political detail. Doing this will expose him to criticism. But this is a risk we must all be prepared to take to break the consensual silence around the ideology of hypercapitalism that is putting human society itself at risk.

Notes

1 Low, N.P. (2020). See especially Harvey, D. (2005) and Streeck, W. (2016).
2 Piketty, T. (2020) pp. 7 and 8.
3 Ibid., p. 1.

4   Ibid., p. 966.
5   Piketty does not provide a succinct definition of hypercapitalism, but the above definition is based on his narrative in Chapter 13 of *Capitalism and Ideology* titled 'Hypercapitalism, Between Modernity and Archaism' and elsewhere in his book. Piketty also frequently uses the term neo-proprietarianism as synonymous with hypercapitalism.
6   Saul, J.R. (2001).
7   Saul, J.R. (1997) pp 20-21.
8   Piketty, T. (2020) p.967.
9   Ibid., p. 970.
10 Ibid., p.991
11 Stiglitz, J. (2023) 'Labor Must Act on Multinationals' *The Age*, 18/07/23 p. 23. Stiglitz, a former chief economist of the World Bank, is co-chair of the Independent Commission for Reform of International Corporate Taxation. https://www.icrict.com/about-icrict (accessed 18/07/2023).
12 Dahl, R. (1985) p. 110. Today, Internet giants such as Amazon, Facebook (Meta), and Twitter are increasingly recognised to be public institutions with enormous potential political power. But they are just the very public face of wider corporate power.
13 To be more specific, if business corporations are predominantly hierarchies whose relationships are organised by the price system, while governments are polyarchies, what does this tell us about the distribution of power in (and therefore the politics of) a society? This question emerges as a subordinate theme in the early pluralism of Dahl and Lindblom, to be taken up as the dominant theme of Dahl's later work, and by Lindblom (1977) in *Politics and Markets*.
14 Piketty, 2020: p. 975
15 Ibid., p. 1017
16 Ibid.
17 Ibid., pp. 1018–1019.
18 Cagé, J. (2020).
19 Piketty (2020) p. 968, notes that 'Some readers may find that the principles of justice I set forth are similar to those formulated by John Rawls' [in Rawls, A *Theory of Justice*, 1971].
20 Piketty (2020), p. 976.
21 Gans, J. and Leigh, A. (2020) p. 3.
22 Ibid., p. 979.
23 According to economists Saez and Zucman (2019) pp 437-511.
24 Piketty (2020), p. 980.
25 Piketty refers to a similar proposal by Anthony Atkinson (2015).
26 Piketty (2020), p. 980.
27 Where income earners pay progressively higher rates of tax the higher the income.
28 https://files.taxfoundation.org/20220418153323/Wealth-Taxes-in-Europe-2022-Wealth-Tax-Countries.png?_gl=1*ko9zad*_ga*MjExMDYyODAyNy4xNjUxNjM4MDky*_ga_FP7KWDV08V*MTY4MTY4NTM2OS4xLjAuMTY4MTY4NTM2OS42MC4wLjA (accessed 26/06/2023).
29 Chancel L., Piketty T., Saez E., and Zucman G. (2022) *World Inequality Report 2022*, p. 138.
30 Ibid., p. 144.
31 Ibid., p. 139.
32 Ibid., p. 140.
33 Ibid., p. 26.
34 Piketty (2020), p. 983.
35 Blanchard is a former chief economist of the International Monetary Fund and Emeritus Professor at MIT. The quote is from the Monetary Policy Institute blog: https://

medium.com/@monetarypolicyinstitute/olivier-blanchard-and-inflation-219f195125fe (accessed 16/04/2023).

36 In Australia, calculated by the ABS Consumer Price Index. Wright, S. (2023) 'Catch-22; RBA faces policy paradox' Melbourne, *The Age*, 30/05/2023 p. 23.

37 Piketty, T. (2022) p. 240.

38 https://www.google.com/search?q=The+Central+Bank+of+the+Argentine+Republic +interest+rate+2023&rlz=1C1GGRV_enAU754AU754&oq=The+Central+Bank+of +the+Argentine+Republic+interest+rate+2023&aqs=chrome..69i57.13059j0j4&sour- ceid=chrome&ie=UTF-8 (accessed 08/04/2023).

39 Piketty, T. (2020) p. 1007.

40 Ibid., p. 1011.

41 Ibid., p. 1011.

42 Based on the OECD Skills Outlook Report 2013. https://www.abs.gov.au/statistics/ people/education/programme-international-assessment-adult-competencies-australia/ latest-release#data-download (accessed 13/04/2023).

43 Perry, L. (2018).

44 Ibid., p.63.

45 Piketty (2020), p. 669.

46 Ibid., p. 1005.

47 Ibid., p. 1005.

48 United Nations (2015) *The Paris Agreement*, https://unfccc.int/process-and-meetings /the-paris-agreement?gclid=CjwKCAjw3POhBhBQEiwAqTCuBqAL6yqkrXSRGuh f66u99FNosDpGitTn24eM15YRliWrLr2LO4IN6BoCmdoQAvD_BwE   (accessed 17/04/2023).

49 Chancel, L., Bothe, P and Voituriez, T. (2023) p. 89.

50 Ibid., p. 89.

51 Ibid. p. 129.

52 Ibid. p. 94. The reference cited is to Naran, B., Connolly, P., Rosane, D., Wignarajah, G., Wakaba, G., and B.uchner, B (2022).

53 Reported by the ABC (https://www.abc.net.au/news/2023-04-19/australias-energy -transition-needs-gas-safety-net-report-finds/102236352). 'The Net-Zero Australia Study will identify plausible pathways and detailed infrastructure requirements by which Australia can transition to net zero emissions and be a major exporter of low emission energy and products, by 2050'. https://www.netzeroaustralia.net.au/wp-con- tent/uploads/2022/08/Net-Zero-Australia-interim-results-draft-public-version-30-Au gust-22.pdf (accessed 18/04/2023).

54 *Climate Inequality Report,* p. 87.

55 Ibid., p. 93.

56 Ibid., p. 87.

57 Ibid., p. 6.

58 Ibid., p. 113.

59 Ibid., p.87.

60 Naran, B., Connolly, P., Rosane, D., Wignarajah, G., Wakaba, G., and B.uchner, B (2022).

61 Songwe, V., Stern, N and Bhattacharya, A. (2022).

62 Ibid., p. 20.

63 Judt, T (2005) p. 67.

64 *World Inequality Report*, Executive Summary, p. 15.

65 Harvey, D. (2005).

66 Ibid., p. 178.

67 Piketty, T. (2020) pp 608-609.

68 Ibid., p. 609.

69 https://www.thebalancemoney.com/national-debt-by-year-compared-to-gdp-and -major-events-3306287 (accessed 28/04/2023).

70 Hayward, D. (2023). Hayward is Emeritus Professor of Public Policy and the Social Economy at the School of Global, Urban and Social Studies, RMIT University, Melbourne.
71 Christopher, B. (2022).
72 Ibid., p. 253.
73 Hayward, D. (2023) citing Christophers B. (2022).
74 Jan-Werner Müller (2022) p. 129, writes that in many political parties 'the number of partisans – as in party members – has been dwindling, while what is sometimes derided as a "blob" of consultants, pollsters and spin doctors has been expanding.'
75 House of Commons Political and Constitutional Reform Committee (2015) *The UK Constitution, A Summary and Proposals for Reform*, p. 5. https://www.parliament.uk /globalassets/documents/commons-committees/political-and-constitutional-reform/ The-UK-Constitution.pdf (accessed 01/05/2023).
76 https://www.law.cornell.edu/wex/constitutional_amendment# (accessed 01/05/2023).
77 Piketty, T. (2020) p. 1017.
78 Ibid., p. 1022.
79 https://www.bbc.com/news/business-58847328 (accessed 27/06/2022).
80 Held D. (1995) p. 135. David Held's conception of cosmopolitan democracy is discussed at greater length in Low NP and Gleeson B. (1998) pp. 183–185.
81 Piketty, T. (2020) p. 1026.
82 Ibid.
83 Ibid., p. 991.
84 Ibid.

References

Atkinson, A. (2015) *Inequality*, Cambridge: Harvard University Press.

Cagé, J. (2020) *The Price of Democracy, How Money Shapes Politics and What To Do About It*, Cambridge: Harvard University Press (tr. P. Cammiler).

Chancel, L., Piketty, T., Saez, E. and Zucman, G. (2022) *World Inequality Report 2022*, World Inequality Lab (https://wid.world/).

Chancel, L, Bothe, P and Voituriez, T. (2023) *Climate Inequality Report, Fair Taxes for a Sustainable Future in the Global South 2023*, World Inequality Lab (https://wid.world/).

Christophers, B. (2022) *Rentier Capitalism. Who Owns the Economy, and Who Pays for It?* London: Verso.

Dahl, R. (1985) *A Preface to Economic Democracy*, Berkeley and Los Angeles: University of California Press.

Gans, J. and Leigh, A. (2020) *Innovation + Equality, How to Create a Future That Is More Startrek than Terminator*, Cambridge: MIT Press.

Harvey, D. (2005) *A Brief History of Neoliberalism*, Oxford and New York: Oxford University Press.

Hayward, D. (2023) 'Back to the future? Cain, Kennett, Andrews and the Victorian budget' (A version of this paper was given at the 2022 Australian Political Studies Association Annual Conference held at the Australian National University).

Held D. (1995) *Democracy and the Global Order*, Stanford: Stanford University Press.

Judt, T (2005) *Post War, A History of Europe Since 1945*, London: Penguin Books

Lindblom, C.E. (1977) *Politics and Markets, The World's Political Economic Systems*, New York: Basic Books.

Low, N.P. and Gleeson B. (1998) *Justice, Society and Nature*, London and New York: Routledge.

Low, N.P. (2020) *Being a Planner in Society, For People, Planet, Place*, Cheltenham: Edward Elgar.

Müller, J.-W. (2022) *Democracy Rules*, Dublin: Penguin Books.

Naran, B., Connolly, P., Rosane, D., Wignarajah, G., Wakaba, G. and Buchner, B. (2022) *Global Landscape of Climate Finance: A Decade of Data*, Climate Policy Initiative. https://www.climatepolicyinitiative.org/wp-content/uploads/2022/10/Global -Landscape-of-Climate-Finance-A-Decade-of-Data.pdf (accessed 17/04/2023).

Perry, L. (2018) 'Educational inequality', in *How Unequal? Insights on Inequality*, Melbourne: Committee for the Economic Development of Australia, pp. 56–66.

Piketty, T. (2020) *Capital and Ideology*, Cambridge: Bellknap Press of Harvard University.

Piketty, T. (2022) *A Brief History of Equality*, Cambridge: Bellknap Press of Harvard University.

Saul, J.R. (1997) *The Unconscious Civilization*, London: Penguin Books.

Saul, J.R. (2001) *On Equilibrium*, Harmondsworth: Penguin Books.

Saez, E. and Zucman, G. (2019) 'Progressive wealth taxation', *Brookings Papers on Economic Activity*. Washington, DC: Brookings Institution.

Songwe, V., Stern, N. and Bhattacharya, A. (2022) 'Finance for climate action: Scaling up investment for climate and development', London School of Economics, Report of the Independent High-Level Expert Group on Climate Finance. https://www.lse.ac.uk/ granthaminstitute/wp-content/uploads/2022/11/IHLEG-Finance-for-Climate-Action-1 .pdf (accessed 18/04/2023).

Streeck, W. (2016) *How Will Capitalism End?* London and New York: Verso.

11

CLIMATE TRANSFORMATION

Action, Actors, and Activists

Introduction

To meet new challenges, modern democracies have transformed societies in the past. They must do so again to meet the challenge of global heating.

We have seen in previous chapters that the world today is facing a threefold interlinked crisis: the slowly evolving climate crisis (Chapter 7), a crisis of democracy and international order (Chapter 8), and a crisis of inequality and poverty (Chapter 9). Transformation requires a change of ideology, meaning a change in the way of conceiving social justice and the distribution of resources and power. An alternative ideology to the neoliberal regression is today taking shape, as we saw in Chapter 10.

Transformation requires political action, activists, and champions to instigate change and make it happen. The full implementation of transformation may occur quite suddenly over a short period of time, but it is preceded by a much longer period in which advocacy of fundamental change builds and builds, step by step until a tipping point is reached. That is the point at which champions of social change at the peak of political power can be most effective. However, history suggests that transformation takes place when there is a widespread feeling in society that something at the core of politics and society has to change. A political ethos is formed that fertilises the possibility of transformation.

In what follows, we will consider the 'Who', 'Where', and 'Under What Conditions' social transformation could occur. From which part of society will the champions of transformation emerge? Who will populate the mass movements that will make transformation a truly democratic movement?

DOI: 10.4324/9781003382133-13

Intellectual Leadership

Today's social democratic movement already has champions among economists such as Joseph Stiglitz and Paul Krugman (Nobel Prize winners) and less well-known stars such as Herman Daly, Steve Keen, John Quiggin, Lucas Chancel, and Thomas Piketty, and Maja Göpel. There is a long list of intellectual champions from science, philosophy, politics, geography, and sociology.

The political economist and Former Secretary General of the German Advisory Council on Global Change, Dr Maja Göpel, in her book *Rethinking Our World* sees the need for fundamental transformation. Following a forensic demolition of the theoretical claims of the neoliberal regression, she writes:

The story of perpetually growing consumption for all has turned out to be a fairy tale, both ecologically and socially. Behind the breathtaking figures, a system has gradually arisen that is destroying our planet, returning ownership structures to those last seen under feudalism, and that relies on constant growth so as not to collapse under the weight of its own inequities.[1]

Internationally, the Secretary General of the United Nations, António Guterres, is a champion of social transformation. But we have yet to see a champion of social democracy emerge from the ranks of national politics. Perhaps, as Piketty remarks, politicians are frightened to move, lest any redistribution of wealth and income would open a 'Pandora's Box'. Thus 'it would be better never to open it than to face the problem of not being able to close it once opened'.[2]

To take that frightening step, politicians need powerful intellectual support. Hayek understood that need in founding the Mont Pelèron Society. The intellectual leadership was to disseminate its ideology through a network of 'second-hand dealers in ideas'. His strategy was effective, but not original since it followed the model of the Fabian Society in Britain. Advice on the return of classical liberalism has since become institutionalised and entrenched.

Like the Mont Pelèron Society, the aim must be to include a range of critical public policy thinking from a broad Left perspective encouraging internal debate. There are institutes around the world with such a perspective. For instance, the Fabian Society (UK), the People's Policy Project (USA), the Australia Institute (Australia), the Broadbent Institute (Canada), TASK (Eire), Demos Helsinki (Finland), the Friedrich Ebert Foundation, the Wuppertal Institute for Climate, Environment and Energy (Germany), and the Foundation for Democratic Reforms (India). There are also institutes with a broad critical focus on national public policy, such as the Grattan Institute (Australia), the Centro de Estudios Públicos (Chile), the Centre of Policy and Legal Reform (Ukraine), the Centre for Fair Political Analysis[3] (Hungary), and the Institute for Social and Economic Analyses (Czech Republic).

Most of these institutes have a national focus, whereas the problem faced today is global or at least international. Perhaps a starting point would be to create a new international forum for social democracy comparable to the Mont Pelèron conference. However, given the hegemony of the neoliberal regression, the power of citizen movements will be needed if transformation is to happen. In this respect, there is a problem.

Nativist Ideologies and the Politics of Identity

There is a seeming madness thrashing about in the world today, desperate for change, hooking itself to the most bizarre ideas imaginable projected on social media. But the people grasping at these ideas are not mad or 'deplorable'. They are looking for fundamental change because they have lost faith in democracies to deliver the sort of future they deserve. But they must not be dismissed and denigrated.

In the absence of a social democratic agenda offering real hope for broad economic and social progress, the working and middle classes of societies have turned to what Piketty terms 'nativist' ideologies. These ideologies turn inward to emphasise nationalism over socialism and even look to fascist solutions.

Nativist ideology can take two forms. In one form, 'social nativism', is strongly aligned with the interests of disadvantaged workers and may advocate national action to promote a measure of social equality among 'people to be considered true natives of the territory in question'.[4] This was the ideology (admittedly racist) of the US Democratic Party in the 1880s. Piketty finds the rise of social nativism in post-Communist Europe and to an extent in India and Brazil.

The other form is 'market nativism' found in Italy in the ideology of the Lega Nord (an anti-tax party) and the Trump presidency.

> In the 2016 presidential campaign, Trump tried to give his politics a social dimension by portraying himself as the champion of the American worker, whom he described as the victim of unfair competition with Mexico and China and as citizens abandoned by Democratic elites.[5]

This stance – generating a trade war with China, building a border wall with Mexico and reducing immigration – was combined with tax cuts for the rich and multinational corporations. In the latter respect, Trump went even further than the Republican presidencies of Ronald Reagan and George H.W. Bush. Though Piketty does not discuss the possibility, there seems to be a tendency for market nativism to encourage anti-democratic forces verging on fascism as illustrated by Trump's refusal to acknowledge defeat in the 2020 election and the subsequent storming of the Capitol.

There has been a change in the social profile of Left voters. Piketty shows how in Europe and the USA less educated voters 'little by little' withdrew their support for parties of the Left. In America, in 2016, these voters supported the Republican candidate, Donald Trump, capturing 46% of the popular vote in 2016, and 47% in 2020. The pattern is supported in Table 11.1 by independent research from the Pew Centre.

Between 1950 and 1970, the vote for various Left parties in Europe was higher among less educated voters. But between 2000 and 2020, the position reversed. It was the more educated voters who supported Left parties. The country examples that Piketty first compares are the USA, Germany, France, Sweden, the UK, and Norway. The magnitude of the change is significant. Comparing the votes for the Left from the most highly educated 10 per cent of voters with the remaining 90 per cent, the difference ranged from 10 per cent to 40 per cent *fewer* in the 1950s to between 0 and 20 per cent *more* in the 2010s. The pattern is repeated in Italy, the Netherlands, Switzerland, Canada, Australia, and New Zealand (from 20 per cent fewer to between 10 per cent and 15 per cent more).

Piketty comments:

> The decomposition of the post-war left-right system and in particular the fact that the disadvantaged classes gradually withdrew their confidence from the parties to which they had given their support in the period 1950–1980 can be explained by the fact that those parties failed to adapt their ideologies and platforms to the new socioeconomic challenges of the past half century. Two of those challenges stand out: the expansion of education and the rise of the global economy.[6]

The question of the politics of race and immigration looms large as a potential explanation of the shift to the nativist Right. Piketty examines the influence of these factors in Europe and the USA. However, he concludes that the educational

TABLE 11.1 United States presidential elections: educational differences amongst voters 2016–2020

Level of education	2016 Democrat Clinton (%)	2016 Republican Trump (%)	2020 Democrat Biden (%)	2020 Republican Trump (%)
Less educated (High school or less)	44	51	41	56
More educated (College graduate)	52	41	56	42
Postgraduate qualification	66	29	67	32

(Source: Pew Research Center, https://www.pewresearch.org/politics/2021/06/30/behind-bidens-2020-victory/pp_2021-06-30_validated-voters_00-03/)

cleavage began to have an effect much earlier than the rising importance of the cleavage around immigration and identity in the 1980s and 1990s. Thus, his analysis leads to the conclusion that racial and immigration factors became a convenient explanation for the plight of disadvantaged workers for parties of the nativist right which opposed government intervention to reduce inequality through fiscal and social reform.

There are two other important features of the political scene common to many countries. One is the rise of 'identity' as a motivating force for mass activism. Identitarian movements overlap nativist movements when identity is associated with a particular country or region. Appeal to the specifically *English* identity seemed to power Brexit. Englishness was portrayed as threatened by all the new workers coming to the UK from the former Eastern bloc countries. Appeal to the Hindu identity helps keep the BJP in power in India, a nation of many regional identities.

But there is also a politics of identity around both race and gender: Black Lives Matter, First Nations movements, and LGBTQI are identity movements demanding recognition and empowerment for sections of society that feel (and often are) powerless. As political scientist Michael Kenny observes,

> The contemporary focus upon the harms and injustices perpetuated on members of unchosen groups, many of whose experiences are also shaped by their vulnerable socio-economic position, might well be seen as a continuation of aspects of the social democratic ambition to make liberalism deliver on its promises of real equality for all[7]

After all, the women's movement has not yet achieved full equality even in the most socially progressive societies.

The Green Movement

A major feature of the past 20 years is the rise in many countries of the environment movement. The Greens, as a parliamentary force, began as a party of the 'grass-roots'. As the Green Movement has developed, Green parties in parliaments around the world have broadened their political platforms to embrace social justice. Internationally, the Greens today offers a comprehensive social-democratic and ecological programme. But Green parties at the national level are also influenced by their grassroots origins and varying ideological nuances. Democratic politics, however, depends on the institution of political parties to resolve such differences in forming policy platforms.

The Australian Greens have been among the oldest and most internationally active of the Green political movements. The world's first Green political party contested elections in Tasmania (an Australian State) in the early 1970s. The Federal Party was founded in 1992.[8] The first international conference of the Global Greens was held in Canberra, Australia's capital. Since then, four other congresses of the Global Greens have been held, the latest in South Korea (2023).

One of the largest parliamentary Greens parties is that of Germany (founded in 1980). The Green Alliance (Bündnis 90) in 2003 held 118 seats out of 736 in the Bundestag (Lower House) and 12 out of 69 in the Bundesrat (Upper House). The German Greens' ideology is described as 'social liberalism'. It endorses social justice and the expansion of civil rights, while economically embracing the 'social market economy'. It views the common good as compatible with individual freedom. From 2010 the Greens in France and Germany in 2010 formed a European coalition of ecological parties called 'Europe Ecology' with presence at the EU level.[9]

The Greens Party of England and Wales has held three seats in Parliament, one in the House of Commons and two in the Lords. However, at the local level the Greens hold over 500 local council seats with three in the London Assembly (as of 2022). As noted in Chapter 8, the Greens parliamentary presence suffers from the 'first past the post' voting convention.

There is a little known Greens Party in the USA which, as of 2022, was the fourth-largest political party in the United States by voter registration. But it holds no seats in Congress or the Senate. The party promotes an ideology of eco-socialism based on the four pillars of ecological wisdom, social justice, grassroots democracy, and non-violence.

The US electoral system always risks splitting the progressive vote, letting the Right into power – as happened when the ecologist Ralph Nader ran for president in 2000. Republican George W. Bush won by one electoral college vote even though the democratic candidate Al Gore won 52 per cent of the popular vote. As in the UK, the Greens cannot succeed in congressional representation without change of the voting convention.

A Greens Party was founded in India in 2021 to ensure that those who depend on the land, forest, rains, and rivers, and on food and fuel security were represented. The Party demands that people's needs are met, and rights respected 'for a just, sustainable and decentralised society with a political-economic model that fits within the Green Philosophy'.[10]

The Global Green political movement has inherited the mantle of social democracy from the Labour movement. The manifesto of the Global Greens boasts, 'We are the only global organisation with 100+ Green Parties world-wide, as well as Members of Parliament, Foundations, Think Tanks and individuals working together for an environmentally and socially just world!'.[11] The Global Greens brings together politicians, members, activists, academics, institutes, the UN, and NGOs from around the world.

The Global Greens Charter was adopted in Canberra in 2001. The foundational principles of the Charter include 'ecological wisdom', 'social justice' and 'participatory democracy'. One of four pillars on which the Australian Greens found their policy reads,

Many of the social problems we have today – crime, discrimination, disease, poverty – could be dramatically improved if we focus on eliminating extreme inequality in Australia and across the world. The Greens believe that it should be the priority of all governments to alleviate poverty and to extend opportunity to all members of society.[12]

In recent years we have seen a rise in disruptive protests, for instance 'School Strike for Climate' initiated by Greta Thunberg, and the 'Extinction Rebellion' actions (traffic disruption and art defacement).

These sporadic disruptions draw media attention. Some Greens supporters believe that provocations are what is needed to keep the problem of global heating and the lack of government action in the public eye.[13] But as the Greens movement begins to occupy the mainstream of politics, as it must, they may not help bring more voters into the Green fold. The Greens need to gain the support of 50 per cent of the population left behind in the neoliberal regression. Destructive protests are designed to annoy people, some of whom may be thinking about the need for social change.

Such distractions divert focus from the main strategy being played by the Greens political parties in democracies – and by the evidence of climate change itself. Nothing comes closer to moving the public's perceptions than the disruptions caused by climate change. Fear works on the public imagination so much more effectively than irritation.

The social transformation necessary to address climate change must reach across a range of social movements, some of them rooted in the misery caused by the neoliberal regression, reaching out even to those now turning away from democracy. Transformational politics must range across classes from the intellectual elites to working people struggling for a better future.

Where Will Transformation Take Hold?

Speculating exactly where and when a national government will be formed, able and willing to implement a social-democratic climate transformation is somewhat fruitless. The present situation is obviously complex. In some ways the present seems to resemble the late nineteenth and early twentieth centuries in Britain when the political Labour movement was poised to replace Liberal governments leaning towards reform.

It is possible that the traditional Social Democrat Left[14] may in time find a way of engaging with and activating the 'less advantaged classes' without alienating the middle and the highly educated. Piketty notes that turnout at elections of the bottom 50 per cent of income distribution has fallen off since the 1990s in France as in the UK and the USA. Piketty argues for a 'social hypothesis', 'that the less advantaged classes came to feel more and more abandoned by the parties of the

Left which increasingly drew their support from other social categories (notably the better educated)'.[15]

It is now more likely that the Green parties will occupy the space vacated by the traditional Left. An independent US think tank, the Council on Foreign Relations, has assessed and mapped the rise of Green parties worldwide.[16] The Council writes that 'green parties are increasingly shaping the debate in countries around the world'.[17] Green parties have evolved 'from single-issue environmentalists into broad-based political parties capable of winning elections and serving at the highest levels of government'.[18]

Green parties are now in governing coalitions in Austria, Belgium, Finland, Germany, Ireland, Luxembourg, and New Zealand. They are represented in national legislatures in Australia, Brazil, Canada, Chile, Colombia, Croatia, Cyprus, France, Italy, Hungary, Latvia, Mexico, the Netherlands, North Macedonia, Norway, Poland, Rwanda, Serbia, Spain, Sweden, Switzerland, and the United Kingdom.

Centre–Left governments, fearing populist Right movements, are beginning cautiously to respond to demands for reform from the Green Left. The 'Green New Deal' is partially incorporated in President Biden's carbon-saving plans for infrastructure, agriculture, and energy. The Labor government in Australia (lacking a majority in the Senate) has had to cede some ground to Green demands for social as well as environmental reform. In Germany, since 2021, the Green Party (Die Grünen) has been influential in policymaking.

As McBride reports, though the tendency is to consolidate around a democratic socialist policy agenda, there is quite a wide range of ideological positions among Green parties worldwide. The proposals of Piketty and his team when they filter through to Green agendas may further consolidate climate policy around social democracy.

In any case, the Greens, like Müller and Piketty (above and in Chapters 8 and 10), agree that pluralism is central to democracy. Many voices will be heard, and varied Green political strategies – even ideological nuances – will evolve in different national settings. In this respect, however, Müller draws attention to the dual nature of democracy:

> First it requires a designated locus (and specific times) for collectively binding decision making – for the expression of political will through law-making: a majority getting its way, after the opposition has had its say. On the other hand, it is in need of a place for the continuous formation of opinions and political judgements in society at large.[19]

Pluralism demands that, within certain limits, we must hear and respect opinions. Müller specifies the limits thus:

Uncertainty – and the exercise of freedom in general – must be contained within two hard borders; people cannot have license to undermine the standing of their fellow citizens as free and equal members of the polity, and while everyone is entitled to their opinions, everyone cannot have their own facts'.[20]

Göpel urges those who seek transformation to 'be kind and patient, but persistent'.[21] If you hit a dead end, 'take a step back and see if there is another possible approach'. Seek allies – 'there are many more around than you may think'. In democracy, 'we rely not only on political actors to have the courage to make difficult decisions; we also depend on the population to support them'.[22]

These features of democracy apply as much within parties as to the public sphere generally. The formation of opinions within parties must occasion open debate, but if the party is to be effective, facts must be brought to bear on policy, and there must be closure: meaning *binding decisions* must be made on policy and strategy. Party members whose voices have been heard but still disagree on details must abide by those decisions.

The path to social democracy is a long one. The groundwork for social democracy is being laid. Once one government adopts a full social democratic programme, as did the UK's Attlee government in 1945, it seems likely, on the basis of past experience, that others will follow. But the Attlee government's programme was conceived during the devastating experience of the Second World War. As discussed in Chapter 3, there was widespread public support for social transformation. This brings us to the final key to the transformative process, the political ethos.

Political Ethos

In Chapter 1, political ethos was defined as 'what is known to be true and what is known to be right' amongst the public at large. Political ethos is about more than ideology and 'includes feelings of justice and injustice deeply embedded in a nation's culture' (from Chapter 3). In Chapter 9, I wrote,

> By 'political ethos' I want to imply something ontological more than sociological: what it is felt to *be* a citizen of a nation, who and how some occupants are excluded from that way of being, and how that sense of being is embedded in the nation's history.

The political ethos in Australia was discussed as an example.

The political philosopher often associated with 'post-modernism' Michel Foucault wrote: 'Finally, all these present struggles revolve around the question: Who are we?'[23] If I focus on the 'who' and the 'we' in this statement it can be read as being about 'identity politics' discussed above. What interests me in this question, though, is not the 'we' but the 'are': the present tense of the verb 'to be': an *ontological* statement.

There is an important distinction between having an opinion and a state of being. The philosopher who probed the question of 'being' most completely was Martin Heidegger. Importantly for a world threatened by climate change, Heidegger argued that our Being is not separable from our surroundings. In his concept of the person as being-there (*Dasein*) he wanted to abolish the idea, from the philosophy of Descartes, of an isolated homunculus in our brain that dispassionately receives and processes sense impressions – essentially the philosophy of individualism.

It is inherent in being human to experience empathy, genuinely feeling the pain of others. But that does not always happen. We might feel for others close to us, but not much for people we don't know. Social philosopher Jeremy Rifkin in his book *The Age of Empathy* wrote that, 'learning how to work together in a thoughtful and compassionate manner is becoming standard operating procedure in a complex, interdependent world'.[24] In our complacent individualistic world that is not happening. But when we all experience an existential threat there emerges a wider sense of empathy.

That occurred, at least in some places, during the Covid-19 pandemic. Governments responded by throwing over individualistic ideologies and helping the poor and disadvantaged. The political ethos shifted, albeit temporarily. During the Second World War the political ethos shifted further and for longer, ushering in social democracy and the welfare state.

The war in Ukraine exemplifies the condition for a similar shift in political ethos as occurred in Britain and Europe during the Second World War. When Putin's war finally ends it is unthinkable that the neoliberal regression and its individualistic ethos, which brought oligarchs to power, will survive. Already the Ukraine government is planning redistributive policies with the broad support of the World Bank.[25]

Climate change is not like modern war. It does not happen suddenly and totally. Global heating is creeping up on us, with sporadic outbreaks worldwide of localised devastations: fires, floods, droughts, and winds. But global heating, once it reaches a certain pitch will probably trigger change in political ethos along the lines of, if we are to care for ourselves, we must care for all those affected, because one day *we* might be affected.

I do not argue that the shift to social democracy is *dependent* on a shift in political ethos resulting from climate change. But it is one important factor, and if, when, and where the political ethos shifts to a more social world view, that will be a trigger for social transformation.

Notes

1 Göpel, M. (2023) p. 83.
2 Piketty, T. (2020) p. 358.
3 Self-described as 'unpopulist' as opposed to 'illiberal'.

4 'Blacks were just as much natives of the United States as whites (and more so than the Irish and Italians)' (Piketty, 2020: p. 245).
5 Piketty, T. (2020) p. 888.
6 Ibid. p. 869
7 Kenny, M. (2004) p.174
8 In the 2022 federal election the Greens boosted their representation in the lower house from one seat to four. They won 12 seats in the Senate, giving them the balance of power.
9 The French website is far from explicit as to the aims and scope of this organisation, https://www.eelv.fr/, (accessed 13/07/2023)
10 https://indiagreensparty.org/2019/12/03/founders-note/ (accessed 04/05/2023).
11 https://globalgreens.org/ (accessed 04/05/2023).
12 https://greens.org.au/about/four-pillars (accessed 04/05/2023).
13 Gibson, S. (2022).
14 in France led by Jean-Luc Mélenchon.
15 Piketty, T. (2020) p. 753.
16 The Council is a nonpartisan nonprofit organization. CFR is based in New York City, with an additional office in Washington. It is chaired by David Mark Rubenstein, an American lawyer and billionaire businessman.
17 McBride, J. (2022). The paper provides a useful chronology of Green party evolution as a political force.
18 Ibid.
19 Müller, J-W. (2022) p. 94
20 Müller, J-W. (2022) p. 184.
21 Göpel, M. (2023) p. 187.
22 Ibid. p. 183.
23 Foucault, M. (1982) p. 212.
24 Rifkin, J. (2009) p. 18.
25 'Social inequality across the regions of Ukraine is low, but the social gap in urban and rural areas remains wide, according to a new World Bank report. Toward a New Social Contract calls for a fundamental rethinking of policies to ease the growing divide between those who benefit from new economic opportunities and those who are left behind in an ever-more flexible economy'. https://www.worldbank.org/en/news/press-release/2018/09/25/in-ukraine-labor-taxation-and-social-policies-must-be-upgraded-to-address-rising-inequality (accessed 09/05/2023).

References

Foucault, M. (1982) 'The subject and power', in H. Dreyfus and P. Rabinow, eds. *Michel Foucault; Beyond Structuralism and Hermeneutics*, Brighton: Harvester Press, pp. 214–232.
Gibson, S. (2022) 'Throwing soup on a Van Gogh and other ways young climate activists are making their voices heard', *The Conversation*, https://theconversation.com/throwing-soup-on-a-van-gogh-and-other-ways-young-climate-activists-are-making-their-voices-heard-193210 (accessed 06/05/2023).
Göpel, M. (2023) *Rethinking Our World, an Invitation to Rescue Our Future*, tr. David Shaw. Melbourne and London: Scribe.
Kenny, M. (2004) *The Politics of Identity, Liberal Political Theory and the Dilemmas of Difference*, Cambridge: Polity Press.
McBride, J. (2022) *How Green-Party Success Is Reshaping Global Politics*, New York: Backgrounder Council on Foreign Relations. https://www.cfr.org/backgrounder/how-green-party-success-reshaping-global-politics (accessed 08/05/2023).

Müller, J.-W. (2022) *Democracy Rules*, Dublin: Penguin Books. Müller is Professor of Politics at Princeton University.

Piketty, T. (2020) *Capital and Ideology*, Cambridge: Bellknap Press of Harvard University.

Rifkin, J. (2009) *The Empathic Civilization, The Race to Global Consciousness in a World in Crisis*, New York: Penguin Books.

12
WHAT WE CAN LEARN FROM THE PAST

If there is one thing, we can learn from Karl Polanyi it is that we cannot understand the present without a long view of the past. Climate scientists take that long view in order to understand climate change, yet the focus of politics and society is the ever-changing present.

It is true that some political leaders have been forced by the ongoing reality of climate change to look to the future. We are told about national targets to reach 'net zero' carbon emissions within 30 years. But the focus on the future can also make us complacent. We are encouraged to believe that technology will save us. But the findings from the science of climate change come thick and fast. Almost every day now a new study is reported that shows us that a devastating future for the planet and for humanity is fast approaching.

Even if, miraculously, the world achieves net zero emissions, global heating will not cease. The heating momentum and its effects will continue for many years afterwards. And what does 'net' really mean? It suggests that market trading will even things out between carbon emitters and carbon savers. But this can only happen if institutions are in place to guarantee that carbon saving measures actually reduce emissions. There are no such guarantees, and without a transformation to re-embed the market in society there never will be. The scale of such institutions must span national, transnational, and even global dimensions.

It is of course true that scientists, entrepreneurs, and businesses, sensing future profits, have responded to the need to develop low carbon energy devices using electricity produced by wind, solar radiation, and other sustainable sources – with battery storage. Their potential for reducing emissions is real, as is the effort made by businesses and consumers. But carbon emissions produced in the whole life cycle of these devices, from mining of materials to disposal of the end product at

DOI: 10.4324/9781003382133-14

the end of its useful life, must be included in calculations of how much carbon emissions are actually reduced and how much collateral damage will be incurred. Under the spell of technological optimism, we do not see such calculations being made.

Our leaders are keener to talk about the future than about what has been achieved in the past 30 years. They do not talk at all about the reality of market society and its effects. They have no interest in systemic institutional change. The truth of the past 30 years is revealed in the figure quoted in Chapter 7. To repeat: atmospheric carbon concentration reached 417.2 parts per million in 2022. Carbon gases in the atmosphere have continued to grow rapidly in an upward curve. It is extremely unlikely that global atmospheric heating will be contained below 2° above pre-industrial times and possible that heating will reach 3° and beyond.

The capitalist system has returned to a condition very like the one Polanyi describes in the period leading up to the two World Wars. The reification of nature and humanity, destabilisation of the global financial system, the erosion of democracy, the rise of fascism, the specious justification of poverty and unemployment, the shocking degree of inequality.

Polanyi was deeply aware of the connection between the global social system and the natural system – including the climate and weather. He writes of the 'vortex of change' of both social and natural systems wrought by market societies. There has been no social transformation that would empower democracies to manage the vortex.

The three transformations described in Chapters 3, 4, and 5 took very different forms. But they should be situated in the history of ideological and political struggles explained by Polanyi's long view of capitalism and its contradictions. There is no final, single, and correct resolution of these contradictions so perhaps we should use the term 'antinomy'[1]: 'a contradiction between two apparently equally valid principles'.[2] In this case, the principle of the market and the principle of environmental and social protection. It is the business of democracy to find a just means of reconciling them. But the means can never be final or perfect. It will be always open to adjustment, revision, and improvement. Democracy allows that to occur.

The transformation in Britain from a society entirely shaped by the market to a society in which the market was embedded took 40 years, with two World Wars, to evolve. In the process modern democracy was born. The process, which occurred and had its roots not only in Britain but across most of Europe, required a concerted effort on the part of intellectual leaders, astute and courageous politicians, and a working-class movement. A 'remarkable consensus' developed spanning between intellectual elites and political leaders and between working and middle classes, more inclusive of both race and gender. 'Faith in the state – as planner, coordinator, facilitator, arbiter, provider, caretaker and guardian – was widespread and crossed almost all political divides'.[3] The role of the state was supported by a political ethos of increased empathy and solidarity amplified after the Second World War.

A question now arises: Why did it take the economic breakdown, the devastation of society, and the mass destruction of two World Wars to turn the voting public towards a compassionate social response? The answer is perhaps that the public were not offered a comprehensive ideological alternative until *after* the devastating events. Today an alternative must be found and implemented encompassing the reality of globalised capital *before* global heating becomes irreversible.

The crisis of the 1970s, though it took various forms, forced governments, lacking an alternative, to return to liberal ideologies from the nineteenth century. Margaret Thatcher's government in Britain started the ball rolling based on a collection of ideologies. These had been well prepared since 1947 by the members of the Mont Pelèrin Society and further refined by monetarism. Thatcher added the discipline of *poverty* to legislation limiting trades union power, recalling the principle evinced by Joseph Townsend in the eighteenth century that the discipline of hunger saves a lot of legislative trouble, is peaceful and silent, and exerts unremitting pressure.

The neoliberal transformation, unlike the social democratic transformation was a 'top down' affair, but it was supported by the British public who were tired of the constant conflict between governments and unions. The middle classes (white-collar public and private employees, small tradesmen and the self-employed) who had thrived under social democracy now felt thwarted in their desire for economic improvement. The fragile consensus around social democracy was broken.

The spread of the neoliberal regression across the world was hugely assisted by Hayek's army of 'second-hand salesmen of ideas' situated in the so-called independent think tanks liberally funded by big business. Eventually these 'independent' sources of advice to government morphed into powerful private global consultancies drawing former public servants into their embrace, further weakening the power of the state to form competent policy.

The transformation from communism to capitalism in Russia and its client states was conducted under the ideology of the neoliberal regression at its most potent. There was an impatience both in the West and in Russia to see the end once and for all of totalitarian communism. The Communist system collapsed under the force of its internal economic failures as much as from the absolute lack of human freedoms. It was simply incapable of delivering what communism promised.

The result of the so-called 'shock therapy' for Russia in the short term was catastrophic. In the long term the humiliation of Russia by the triumphalist rhetoric of the West fed a deep resentment that empowered Putin and his allies to invade Ukraine and threaten the former states of the USSR. A 'lost idea' that consolidates the Russian nation and marks new boundaries between 'us' and 'them'.

While transformations may be devised in detail by intellectuals and implemented from the top by political elites, they are powered and driven by mass movements in civil society. The transformation of market society to social democracy in Britain could not have occurred without the mass movements of working people and the feminists. In Europe's Eastern Block dissent seethed below the surface in the midst of authoritarian communism.

Humanity, Piketty observes, has progressed dramatically in terms of health and education over the last 200 years.[4] But the market has neither limits nor morality, echoing the words of sociologist Max Weber. Human progress is fragile and constantly threatened by inequality and the sort of inadequacies and flaws of democracy discussed in Chapter 8.

The climate crisis seems slow moving – in terms of the politics of today that is focused relentlessly on the present. Other crises seem more immediate: war in Europe and the threat of war in Asia, the possible breakdown of the world economy caused by political brinkmanship in the USA, the potential threat to human civilisation of artificial intelligence, the potential of future global pandemic, the crisis of inequality and poverty both globally and nationally.

Democracies under the neoliberal regression have been unable to meet public demands for high-quality social services, integrity in politics and commerce, and social justice. In addition, there is a growing global movement demanding restitution for the evils of colonialism and slavery which delivered untold wealth to Britain, the USA, and Europe.

All of these crises are connected with the deficiencies of today's ideologies and democratic institutions. How effective can regulation be when the market rules society? Thomas Piketty and the *World Inequality* group take up the narrative of critical history in the steps of Polanyi. They provide a detailed analysis of the problem of inequality and its supportive ideology (Chapters 9 and 10). They offer a template for an alternative social democratic politics.

The ethical vacuum at the heart of hypercapitalism is threatening democratic societies. Under this ideology the justice of the market is all there is. Everyone gets what they deserve, billionaires and the homeless alike. If they get it, they deserve it. The more they get, the more they must deserve it. Everyone in democratic societies feels the injustice in such vacuous assumptions. The danger to democracy is the ethical vacuum: no truth, no social justice, no reality, no facts. Unless the vacuum is filled by a truthful and ethical ideology, it will be filled by the slogans of clownish plutocrats like Donald Trump.

In which nation might democracy embrace a new democratic socialism? Of all the democracies, the United States seems unlikely to do so. Its individualist political ethic, its sclerotic democratic institutions, and its polarised politics seem more likely to lead to revolution than peaceful transformation. Britain, Australia, and New Zealand might take tentative steps towards transformation, as might the European Union. Though the latter would find it hard to risk moving away from the 'ordo-liberal' concept of the 'social market economy'. Ukraine, in the midst of suffering, demonstrating every day an empathic social cohesion born of war, is already taking steps to abandon the neoliberal creed.

Elsewhere everything is unpredictable. But transformation, or as Hall puts it, 'third order change' does not occur from incremental steps. It is a disjunctive process associated with discontinuities in policy.[5] As climate change forces adaptation of humanity in different regions of the globe, social norms, political ethos, and institutions may shift suddenly.

Chapters 10 and 11 postulate a 'practical utopia'. The agenda set out by Piketty and his colleagues could perhaps be construed as post-modern socialism, though the term 'post-modern' now seems dated. The sociologist Peter Beilharz, writing on the eve of the twenty-first century, remarked, 'Modernity, after all, is always reforming itself, and the logic of the system is routinely violated or balanced out in its local regions or parts'. Modernity's hallmark, with its Enlightenment ethics, Beilharz continues, is a sense of crisis, failure, dissatisfaction. 'Encountering the post-war boom in affluent societies, we imagined that it would last forever, mesmerised as we were by the freeze-frame of full employment and the consumer revolution. But now social amnesia takes it revenge on us'.[6]

We need to revisit our history. We need to hope and believe that human institutions can adapt as the biosphere itself adapts to what humanity has forced upon it. We need to protest, lobby, and above all vote to make that adaptive transformation happen.

Notes

1 In the philosophy both of Immanuel Kant and Max Weber.
2 Merriam Webster dictionary definition.
3 Judt T. (2005) p. 362.
4 Piketty (2020) p. 16.
5 Hall, P. A. (1993) pp. 275–294.
6 Beilharz, P. (1994) p. 55.

References

1. Beilharz, P. (1994) *Post-Modern Socialism, Romanticism, City and State*, Melbourne: Melbourne University Press, p. 55.
2. Hall, P.A. (1993) 'Policy paradigms, social learning and the state: The case of economic policy-making in Britain', *Comparative Politics* 25(3): 275–294.
3. Judt, T. (2005) *Post War, A History of Europe Since 1945*, London: Penguin Books.

BIBLIOGRAPHY: KEY READINGS

Addison, P. (1985) *Now the War Is Over, A Social History of Britain 1945–51*, London: Jonathan Cape and the BBC.

Agyeman, J., Bullard, B. and Evans, B. eds. (2003) *Just Sustainabilities, Development in an Unequal World*, London: Earthscan Publications.

Carter, F.W. and Turnock, D. eds (1993) *Environmental Problems in Eastern Europe*, London and New York: Routledge.

Chancel, L., Piketty, T., Saez, E. and Zucman, G. (2022) *World Inequality Report 2022*. https://wir2022.wid.world/.

Chancel, L., Bothe, P. and Voituriez, T. (2023) *Climate Inequality Report, Fair Taxes for the Global South 2023*, World Inequality Lab Study 2023. https://wid.world/wp-content/uploads/2023/01/CBV2023-ClimateInequalityReport-2.pdf.

Daly, H. (1996) *Beyond Growth, The Economics of Sustainable Development*, Boston: Beacon Press.

Göpel, M. (2023) *Rethinking Our World, an Invitation to Rescue Our Future*, tr. David Shaw, Melbourne and London: Scribe.

Hall, P.A. (1986) *Governing the Economy, The Politics of State Intervention in Britain and France*, Cambridge: Polity Press.

Harvey, D. (2005) *A Brief History of Neoliberalism*, Oxford and New York: Oxford University Press.

Hayek, F. (2005) *The Road to Serfdom with The Intellectuals and Socialism*, London: Institute of Economic Affairs.

Judt, T. (2005) *Post War, A history of Europe Since 1945*, London: Penguin Books.

Low, N.P. (2020) *Being a Planner in Society, For People, Planet, Place*, Cheltenham: Edward Elgar.

Low, N. and Gleeson, B. (1998) *Justice, Society and Nature, An Exploration of Political Ecology*, London and New York: Routledge.

Lynas, M. (2008) *Six Degrees, Our Future on a Hotter Planet*, Washington: National Geographic.

Müller, J.-W. (2022) *Democracy Rules*, Dublin: Penguin Books. Müller is Professor of Politics at Princeton University.

Piketty, T. (2014) *Capital in the 21st Century*, Cambridge: Bellknap Press of Harvard University.

Piketty, T. (2020) *Capital and Ideology*, tr. Arthur Goldhammer, Cambridge: Bellknap Press of Harvard University.

Piketty, T. (2022) *A Brief History of Equality*, tr. Steven Rendall, Cambridge: Bellknap Press of Harvard University.

Thompson, J. (2009) *Intergenerational Justice: Rights and Responsibilities in an Intergenerational Polity*, London and New York: Routledge.

Thunberg, G. ed. (2022) *The Climate Book*, London and New York: Allen Lane/Penguin/Random House.

Turner, R.S. (2008) *Neo-liberal Ideology, History, Concepts and Politics*, Edinburgh: Edinburgh University Press.

INDEX

Printed in the United States
by Baker & Taylor Publisher Services